CHRISTINA HARRISON
TONY KIRKHAM

Remarkable
TREES

The University of Chicago Press

HALF-TITLE The strange dragon's blood tree
of the island of Socotra.
FRONTISPIECE Chestnut-leaved oak (detail),
by Masumi Yamanaka.
CONTENTS PAGES Page 5: Nutmeg; hazelnut;
mulberry. Page 6: Douglas fir; oak; pecan.
Page 7: Sweet chestnut.

The University of Chicago Press, Chicago 60637

Remarkable Trees
© 2019 Thames & Hudson Ltd, London
Text and illustrations © 2019 the Board of
Trustees of the Royal Botanic Gardens, Kew,
unless otherwise stated, see p. 252

Designed by Lisa Ifsits

Published 2019

28 27 26 25 24 23 22 21 20 19   1 2 3 4 5

ISBN-13: 978-0-226-67391-2 (cloth)

First published in the United Kingdom in 2019
by Thames & Hudson Ltd, 181A High Holborn,
London WC1V 7QX. Published by arrangement
with Thames & Hudson Ltd, London.

Library of Congress Cataloging-in-
Publication Data
Names: Harrison, Christina (Horticulturist),
author. | Kirkham, Tony, author.
Title: Remarkable trees / Christina Harrison
and Tony Kirkham.
Description: Chicago : The University of
Chicago Press, 2019. | Includes bibliographical
references and index.
Identifiers: LCCN 2019008169 |
ISBN 9780226673912 (cloth)
Subjects: LCSH: Trees.
Classification: LCC SD383 .H65 2019 |
DDC 582.16—dc23
LC record available at
https://lccn.loc.gov/2019008169

Printed and bound in China
by C & C Offset Printing Co. Ltd

CONTENTS

Introduction 8

BUILDING
AND CREATING 14

FEASTING AND
CELEBRATING 66

HEALERS
AND KILLERS 114

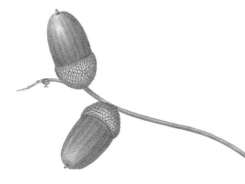

Eichbaum.

INTRODUCTION

Trees have long been important to us, not just for their natural beauty and character, but also because through the ages they have been central to our existence in numerous ways. We have lived with them and among them for millennia and they continue to feed us, shelter us and inspire us. They supply many of the vital ingredients for life – food, medicine, timber, oils, resins and spices – and also ecological services such as providing the oxygen we breathe, controlling soil erosion, trapping pollution, acting as carbon sinks and increasing water purity, as well as moderating the climate. Scientists often refer to these benefits as 'natural capital'. In addition to providing such benefits, trees have also become the subjects of song, poetry, stories and art, and they are woven into our religions, folklore and customs. They act as direct links to our history and can exert a powerful influence over our imaginations and memories.

The commonest botanical description of a tree is a plant that has a self-supporting perennial woody stem. It does not have to be tall or reach a certain age, and indeed some trees grow as shrubs or as dwarfed species and some are quite short-lived. The definition of a tree is surprisingly open to interpretation, and varying descriptions can be found – hence the variations sometimes quoted for the total number of tree species in the world. Here, we use a wider definition of what constitutes a tree and have included some palms too, even though they do not produce secondary growth in their stems (they are monocotyledons), and so could not be called 'woody'. However, self-supporting perennial species such as the coconut fill very similar niches to woody trees in the habitats in which they grow and we wanted to include their fascinating stories.

Trees are a very diverse group of plants, found around the world in an amazing variety of forms and sizes and growing in a great range of habitats. One estimate is that there are around 60,000 species, from

Throughout history we have found both natural beauty and utility in the trees around us – they have not only been used to provide us with shelter and keep us warm, but have also supplied us and our animals with a wide variety of food. In the past, pigs were often allowed to forage for fallen acorns in oak forests, as they still are in places today.

diminutive sub-arctic tundra types including birches to towering tropical hardwoods like mahogany, and from the dragon's blood tree growing in the arid landscapes of Socotra to mangroves that live happily in saltwater. Trees are remarkable examples of evolution, having developed over the past 360 million years to suit different climates, soils, rainfall patterns and any niche they can fit into. This evolutionary drive often results in special adaptations – from resins and saps that deter predators or heal wounds, to attractive fruits that encourage seed distribution – the very things that make them useful to us too.

Symbols of stability, majesty and longevity, trees are some of the oldest, largest and most impressive organisms on the planet, and seem to exist on a different timescale from us.

Too often trees can be seen as commonplace, a green backdrop to our lives, but they should never be taken for granted. A visit to any arboretum or botanic garden such as the Royal Botanic Gardens, Kew, is as if you are walking through the pages of a good book that you can't put down, with every tree a page in that book, and every page telling a different and fascinating story. In this book, we want to bring you closer to over sixty special trees, representing most of the world's major habitats – just some of the species that could be considered remarkable. Trees have also inspired artists, explorers and botanists alike for centuries, and these pages are illustrated with some of the best artworks of tree species from the library and vast collections of the Royal Botanic Gardens, Kew.

ABOVE The sago palm provides a staple element in the diet of many people across Malaysia and Indonesia. It is thought that it is one of the earliest crops to have been exploited and used as a food in the tropics.

OPPOSITE Nutmegs were once so valued that they changed the course of human history. Thousands of people lost their lives in the fight to control the nutmeg trade, and these fruits were once worth more than their weight in gold.

The value and importance of trees can be seen in so many ways. We build and create with an enormous variety of timbers, have discovered which parts of trees taste delicious, and which can kill or cure us, and which species add colour and spirituality to our lives. Many of these trees have changed the course of history and have added to our culture, economy and society. All have intriguing stories to tell. For instance, did you know that nutmegs were once worth more than their weight in gold? And do you know where wild frankincense trees still grow, and how this precious resin, once the world's first Christmas present, is still harvested? Or which tree has the largest fruit in the world; which can give you a serious headache just by sitting under it; and which can kill you in three hours? The answers can be found in this book.

Symbols of stability, majesty and longevity, trees are some of the oldest, largest and most impressive living organisms on the planet and seem to exist on a different timescale from us. Within these pages we will take you on a tour of some of the most remarkable tree species of the world. The yew and bristlecone pine number among the oldest known living things, while the redwoods and eucalypts tower above all other trees at record-breaking heights. Quinine, chocolate, olive and ebony are highly valued as some of the most desirable trees for their products, while others continue to be revered – including the banyan and baobab.

OPPOSITE The Indian banyan fig is a giant of the botanical world. It is also considered sacred in the Hindu religion and is a symbol of faith in India.

BELOW Trees are among the largest living things on Earth. Coast and giant redwoods in California, USA, are some of the most impressive and beautiful trees on the planet.

It is estimated that over 8,000 tree species are under threat of extinction and are precariously balanced in terms of survival. Any species becoming extinct is a tragedy as it represents the loss of an integral piece of the complex web of life in the habitat in which it has evolved. Trees help us thrive and survive, but they are also pillars of stable ecosystems and part of vast ecological networks that support many other living creatures – without trees many of them, from insects, birds and mammals to fungi and bacteria, would completely fail.

We hope this book inspires you to appreciate the wonder and importance of trees and the incredibly diverse range found around the world. In the stories revealed here it is obvious that trees enrich us, and also why we should care about them – they are an intimate part of our lives and our culture, our past and our future.

Cedar, *Cedrus libani*

BUILDING AND CREATING

From the mighty oak to the strong yew and the versatile hazel, throughout human history trees have provided us with materials for building and creating. We have constructed humble dwellings, magnificent halls and imposing temples from their timber, as well as making fencing, baskets, furniture, canoes and wagons, all of which have made our lives easier. Wood was indispensable for ships and weapons, which allowed civilizations to expand, and for trading with and conquering new lands. The paper made from trees has been used to create the scrolls, books, drawings and paintings that have enlightened the world.

Hard-won experience of timbers from different species of tree, and knowledge of their multitude of uses, has been passed down over the centuries, so that even today we know that oak is durable when building a home, sweet chestnut is good for fence posts and barrels, and the black locust makes long-lasting flooring. The Sitka spruce is perhaps one of the most adaptable timbers, used for everything from hats and rope to pianos and guitars, boats and aircraft.

Yew was once the timber of choice for the longbow in Britain, and many other trees have also traditionally been valued for special purposes and for their symbolism. Cedars were highly prized in antiquity for building temples and ships, such as that buried next to the Great Pyramid in Egypt. In Japan the foxglove tree is used to make fine musical instruments and dowry chests. Mahogany furniture and inlays were once a sign of status and wealth. Perhaps the most specific use of a tree is the particular type of willow favoured for making the best cricket bats. And it is not only the timber of trees that we exploit – the bark of the cork oak has been used in many different ways, including for the all-important bottle stopper.

Previous generations understood the importance of cultivating woodlands for sustainable crops of trees, and how, if well-managed, they could be a great resource and the sign of the prosperity of a community. In recent times, with other, synthetic materials available, we have ceased to rely so directly on trees, and have lost some of our deep connection with them. However, we do not have to look far to appreciate the importance of trees to our existence. We continue to build homes and make furniture, and create books, sculptures and objects of great beauty from timber. It is still possible to see and even touch a tradition stretching back many thousands of years, and so re-establish our relationship with trees.

Cedar

Cedrus libani, C. deodara and *C. atlantica*

A majestic architectural tree with distinctive tiers of horizontal branches spreading wide and reaching out into space, *Cedrus libani*, the cedar of Lebanon, has long had a special place in human history, stretching back to perhaps the earliest literature in the world. The Epic of Gilgamesh is an ancient Sumerian poem written around 2000 BC, which tells the story of the hero-king Gilgamesh. With his companion Enkidu, he journeys to the forest of cedars to fight the monster Humbaba; having killed him, they then cut down many of the trees.

The cedar of Lebanon appears numerous times in the Bible as a symbol of might and beauty. It was also thought to have medicinal properties: in the Old Testament, God ordained to Moses that cedar wood was one of the ingredients priests should use to treat leprosy. Because of the strength, fragrance, durability and resistance to insect infestation of the beautiful fine-grained wood, it was exploited by many ancient peoples, including the Phoenicians, Israelites, Egyptians, Babylonians, Persians and Romans. King Solomon sent to Lebanon for cedar timber to build his Temple in Jerusalem. The Phoenicians became the first sea-trading nation in the world by making their merchant ships from cedar, and the boat buried next to the Great Pyramid of Khufu at Giza, dating to around 2500 BC, was built of huge pieces of this wood. The Egyptians also used cedar resin in the mummification process. Later, the Roman author Vitruvius stated that the roof of the Temple of Diana at Ephesus, one of the seven wonders of the ancient world, was built of cedar.

This imposing evergreen conifer, which can grow to 35 metres (115 feet) tall, is native to the eastern Mediterranean, including the Taurus Mountains of Turkey, Syria and Lebanon, where the largest trees are said to be several thousand years old. In its homeland of Lebanon, the natural populations are in decline and under threat from

The oval cones of the cedar of Lebanon develop at the tips of the branches, often only every other year. When ripe they shatter, releasing the seeds inside.

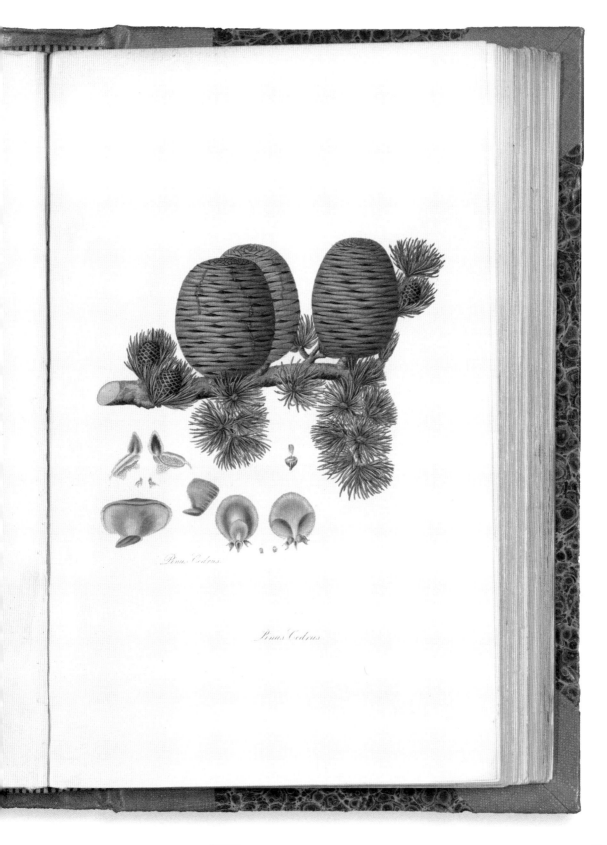

Pinus Cedrus

Pinus Cedrus

CEDRVS *foliis rigidis acuminatis non deciduis, conis subrotundis erectis* Plant. fol. tab 1.

a. juli gema; b. julus cum calyce pronascens; c. idem per longitudinis medium dissectus; d. in magnitudine aucta; e. julus perfectus; f. ejus calyx persistens; g. calyx separatus a facie interiori; h. julus separatus et per longitudinis medium dissectus; i. ejus axis; k. stamina aliquot separata; l. stamen in magnitudine aucta cum filamento m. per-brevi, anthera n. magna et o. ejus extremitatis squama; p. anthera transversè dissecta; q. julus aridus; r. stamē a facie superiori, s. in magnitudine aucta, t. a facie laterali, u. inferior, x. in magnitudine aucta, y. pars pollinis quem continet; 1. conus junior, 2. ejus calyx; 3. calyx separatus a facie externa, 4. interna, 5. ejusmodi conus per longitudinis-medium dissectus, 6. ejus axis, 7. ejus pars inferior, 8. axis denudatus, 9. squamula separata a facie inferior, 10. superior; 11. conus paullo adultior, 12. coni maturi pars inferior, cujus squamis semina incumbunt, 13. ejus axis; 14. squamarum una separata cum binis seminibus, 15. bina unius squamæ semina separata.

past over-exploitation of the timber and damage from insect pests, grazing goats and modern winter sports activities. The last remaining groves growing on Mount Lebanon are known as the 'Cedars of God' and were added to the UNESCO list of World Heritage Sites in 1998 to afford them the protection they deserve.

The exact date of the introduction of cedar of Lebanon into Europe is uncertain. In the British Isles, Edward Pococke, a scholar of Arabic at Oxford University and chaplain in the Holy Land, is said to have grown a tree from a seed that he brought back from Syria in 1638, which still survives today in Oxfordshire. Cedar trees have since been grown as ornamental specimens, along with other exotic trees, in many large parks and gardens throughout Europe and in the United States. They were used to great and lasting effect by eminent landscape architects including William Kent and Lancelot 'Capability' Brown.

The cedar of Lebanon appears numerous times in the Bible as a symbol of might and beauty.

There are actually three other species of true cedars that can be found in the northern hemisphere. All are similar in having stately horizontal branches, blue-green needle-like leaves and oval cones that shatter to release the seeds. *Cedrus atlantica*, the Atlas cedar, grows in the Atlas Mountains of Morocco and Algeria. *Cedrus deodara*, the deodar cedar, is native to the Western Himalaya; both the botanical and common name share the word *deodara*, which means 'timber of the gods' and is derived from the Sanskrit *devadāru* ('deva', god or divine, and 'daru', wood or tree). The deodar is the national tree of Pakistan, and as with the cedar of Lebanon, its timber was in great demand for building temples because of its resistance to decay and close grain that polishes well. The famous houseboats of Srinagar in Kashmir are built from deodar cedar wood, and because of its anti-fungal and insect-repellent properties it was used to make meat and grain stores in the Shimla, Kullu and Kinnaur districts of Himachal Pradesh in India.

The cedar of Lebanon is now under threat in its homeland of the eastern Mediterranean. However, from the eighteenth century onwards cedars were often planted in large gardens and parks in Europe, usually as specimen trees in order to show off their graceful sweeping branches and dark foliage, and many of these trees still survive.

The third species is the rare Cyprus cedar, *Cedrus brevifolia*, which is endemic (it grows naturally only there) to a small area in the Troödos Mountains of central Cyprus. As a result of this restricted island distribution, this cedar is vulnerable and near critically endangered, under threat from fire and climate change.

The true cedar is not to be confused with several other trees that are often given the same common name, such as the Western red cedar and the incense cedar, which are not true cedars at all, but two completely different genera, *Thuja* and *Calocedrus* respectively. The pencil cedar is also a different taxon, *Juniperus virginiana*; as its common name

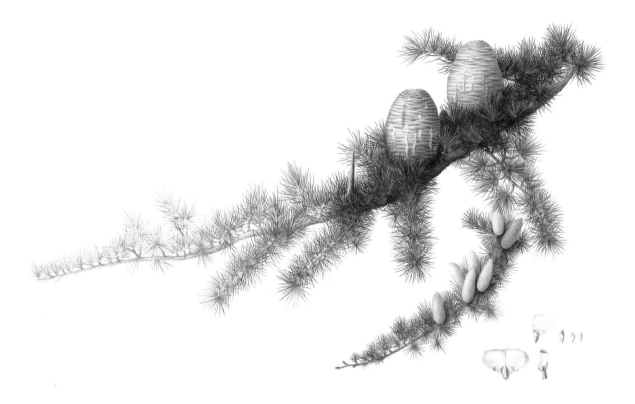

The dark green needle-like leaves of the cedar form in clusters along the branches. Female flowers develop into a green cone which ripens to brown. Separate male cones produce pollen.

suggests, its wood was once regarded as one of the best for making pencils. It was called a cedar because of the similarity of its timber to the aromatic pinkish-brown heartwood of the true cedars.

In the early second century AD, cedars of Lebanon were so prized by the Roman emperor Hadrian that he set up stone boundary markers in the eastern Mediterranean to create an imperial reserve for the remaining trees, hoping to protect them and prevent more being cut down. Sadly, since 2013, the cedars, particularly the Atlas and deodars growing in parks and gardens, have faced a new threat from a disease caused by a fungus (*Sirococcus tsugae*) that causes a pink discolouration of the needles followed by browning and dieback in the tops of the trees. Climate change, which is affecting the moist and cool conditions the cedars of Lebanon need to thrive in their natural habitats, is also threatening these trees. Efforts are being made to regenerate the cedar forests in Lebanon to ensure the survival of these stately trees of great antiquity, which have seen so many human civilizations come and go.

Mahogany

Swietenia macrophylla

If there's one tree that is associated with the finest furniture, albeit from an almost bygone age, it's mahogany (*Swietenia*). Now more often seen as antiques in museums and country houses than in ordinary homes, fine mahogany tables, chairs, cabinets and panelling were once highly sought after. (Indeed, Chippendale mahogany furniture is still highly prized and costly.) Mahogany's ruby-red timber glows when polished, and this lustre, together with the wood's close grain, ensures any object crafted from it has a warm and tactile quality.

This enormous and long-lived tree can soar above the rainforest canopy in its native Central and South America.

Many musical instruments were also once made from mahogany – it is known as a 'tonewood' as its dense nature helps to produce a mellow and balanced tone from the instrument. It has been used for guitar backs, mandolins, drums and expensive violins, but is much less likely to be found in instruments today as the tree is now endangered – a direct result of its popularity and sought-after qualities.

The most important commercial species is *Swietenia macrophylla* – the big-leaved mahogany, also known as 'caoba' in Latin America. Growing to around 30–40 metres (100–130 feet) tall, and sometimes reaching 60 metres (almost 200 feet), this enormous and long-lived tree can soar above the rainforest canopy in its native Central and South America. Its leaves can reach up to 50 centimetres (20 inches) long, hence its name *macrophylla*, or 'big-leaves'. Found sparsely dotted within mixed lowland tropical forests, often near rivers or in deep soils, this mahogany plays a significant role in local culture and the ecosystems of many other species including the endangered giant river otter. The trees help to maintain the stability of the riverside soils where they grow, reducing erosion and helping areas recover from flooding.

OPPOSITE The first mahogany to be discovered and exploited was the Caribbean mahogany. Its richly coloured wood became so highly sought after that it is now considered commercially extinct.

RIGHT Mahogany species have a long history of being felled for their timber. However, if left to regenerate naturally some populations could start to recover.

Big-leaved mahogany is one of the most commercially valuable trees in the Amazon and can be sustainably harvested if strict management plans are followed to avoid over-exploitation. However, although it is now protected by law under CITES (the Convention on the International Trade in Endangered Species), much illegal logging still takes place, affecting tree populations, regeneration and the natural habitats where it grows. This particular species is today more widespread than the other two closely related mahoganies, but its numbers are thought have decreased by 70 per cent since the 1950s, and it is under threat from both logging and deforestation for agriculture.

Swietenia mahagoni, the Caribbean mahogany, was the first mahogany to be discovered by Europeans and was exported in the seventeenth and eighteenth centuries. Its wood is regarded as the most attractive and it was completely over-exploited. Both it and *Swietenia humilis*, the Honduras mahogany, are now considered to be commercially extinct. Over the years, attempts have been made to create plantations of big-leaved mahogany (notably in India), but with limited success because of its vulnerability to the larvae of the shoot-borer insect.

Mahoganies produce separate male and female flowers, which are pollinated by bees and thrips. The woody fruits are packed with seeds that have a tail-like wing so they can rotate and fly on the breeze up to 500 metres (1,650 feet) from the parent tree. If left to grow naturally, mahoganies are pioneer species and can regenerate successfully on disturbed ground or in gaps in the canopy. But their timber and other products usually prove too much of a lure for commercial interests to allow them to reach maturity and achieve their full potential as a majestic and beautiful, as well as an extremely useful, tropical tree.

Black locust tree

Robinia pseudoacacia

Native to North America, the black locust tree is now naturalized in many temperate regions of the world. It is thought that it only occurs naturally in two relatively small, separate areas of the eastern and southern United States – the Appalachian Mountains south of Pennsylvania and the Ozark plateau west of the Mississippi. However, it is now widely cultivated across the continent and can be found growing in 48 states, eastern Canada and British Columbia. It has become successful as an invasive weed tree in woodlands, along railways and on roadside verges because of the abundant amount of viable seeds it produces each year and its ability to regenerate underground prolifically from root suckers, taking advantage by invading cleared woodlands and disturbed forest edges in sunny positions.

Honey from this tree is floral, delicate ... and regarded as one of the most desirable by the connoisseur.

The wood of the black locust is one of the hardest in North America, competing with hickory, but is more resistant to rot, making it a popular timber. It was once used for cart wheels and is prized for furniture, flooring and split post-and-rail fencing, as well as for the construction of rustic playground equipment. Once the bark is stripped, the exposed yellow-coloured wood is smoothed and finished with oil. Because of its durable properties, the black locust was also an important material in the shipbuilding industry, including for making pegs or 'treenails' used to fasten timbers together.

The black locust was first introduced to Europe in 1601, and seeds were planted in a square near Notre Dame in Paris by Jean Robin, who was botanist and herbalist to Henry IV of France at Fontainebleau. A second specimen propagated from cuttings was planted in 1636 in the Jardin des Plantes by Jean's son, Vespasien, who was also a royal gardener to the king. The genus *Robinia* was later named in honour

Robinia was named by Carl Linnaeus after Jean Robin and his son Vespasien, who were gardeners to the king of France and planted specimens of the tree in Paris. It is known as the black locust tree in America, possibly because of its dark bark and its seeds, which resemble those of the locust tree or carob.

of father and son by the Swedish botanist Carl Linnaeus. The species name *pseudoacacia* literally translates as false acacia, one of the common names for this tree since it is not a true acacia. It was introduced to Britain in 1636, and has become a popular tree for planting in urban landscapes because of its tolerance of industrial pollution. An early planting in Britain is now one of the 'Old Lions' of the Royal Botanic Gardens, Kew – the oldest trees of known date still in the gardens. It can be found growing in the original 2-hectare (5-acre) arboretum planted by Princess Augusta in 1762. Although a majority of its trunk now consists of dead wood held together with metal bands, the tree still enjoys rude health and continues to grow happily.

In its early years, the black locust is a fast-growing tree, and can ultimately reach a height of around 30 metres (100 feet). Over time its trunk becomes contorted and deeply furrowed, taking on a very rugged, oriental appearance with lots of character. The young shoots and twigs are armed with pairs of exceptionally sharp woody spines at the base of the grey-green feathery leaves. The individual leaflets of these leaves fold together in rain and also change position at night, a process known as 'nyctinasty', a typical habit of many members of this family

(Fabaceae). One of the best attributes of the black locust are its large hanging clusters of showy white flowers. They are highly scented and have a low pollen count and plentiful nectar, making them attractive to bees. 'Acacia' honey from this tree is floral, delicate, almost transparent and regarded as one of the most desirable of honeys by the connoisseur. It is readily absorbed by the body and is much richer in fruit sugar than other varieties, so has a low glycaemic index. In Italy fresh black locust flowers are collected, dipped in a light batter and deep-fried to make delicate, sweet fritters known as 'frittelle di acacia'. The flowers are also used in perfumery.

No other American tree has been so successful at making its home in the woodlands and gardens of Europe as the black locust. However, despite the fact that William Cobbett, farmer and famous radical politician, extolled its qualities – he called it 'the tree of trees' in his 1825 book on silviculture, *The Woodlands* – it never became popular as a forestry tree. Cobbett set himself up as an importer and dealer in seeds and plants, saying that he had sold over a million trees and was unable to meet demand. Though its timber is useful, the black locust refuses to grow straight, so it was rarely planted except as an ornamental tree in parks and gardens, simply for pure pleasure.

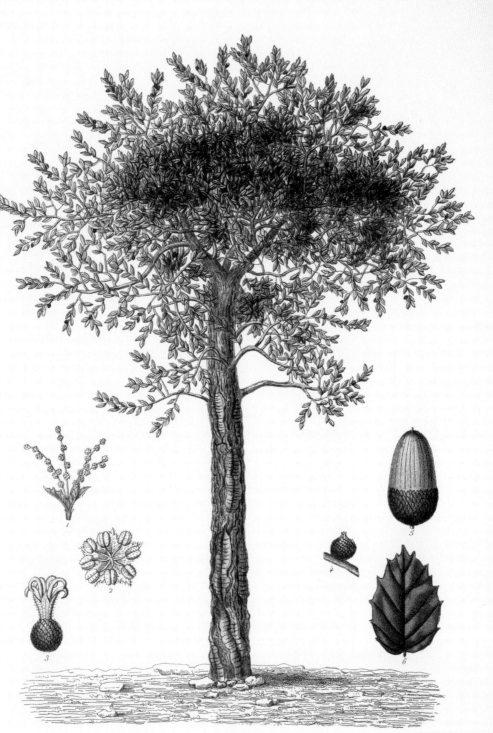

Cork oak

Quercus suber

At the ripe old age of 234 years, the oldest and largest cork oak tree in the world can be found growing in the Portuguese town of Àguas de Moura. Known as the 'whistler cork oak', from the sounds of the hundreds of songbirds that roost in its branches, it was made a national monument in 1988. It measures over 16 metres (52 feet) high, and five people with linked arms are needed to surround its trunk. This phenomenon of a tree has been stripped for cork at least twenty times, and in 1991 was famous for generating 1,200 kilograms (2,645 pounds) of bark, which produced over 10,000 corks, more than the average cork oak tree would produce in its entire lifetime.

Our familiarity with the many useful attributes of cork is nothing new. Etruscan amphoras recovered from a shipwreck off Grand Ribaud, France, dating to the sixth or fifth century BC, were carefully stoppered with cork. Later, in the first century AD, the Roman author Pliny the Elder referred to cork oak trees in his celebrated work *Natural History*, explaining how the bark was used as floats for anchor ropes and fishing nets, as bungs for casks and as the soles for women's winter shoes. Another Roman author, Columella, writing on agriculture, recommended making beehives from cork where available, as it maintained an even temperature.

The cork oak is a slow-growing but long-lived, medium-sized evergreen tree, up to 20 metres (66 feet) tall, with a broad, spreading crown, often wider than the tree is high. Individuals can live up to 200 years on average. It grows naturally on the Iberian Peninsula, in Portugal and Spain, as well as the coastal regions of northwest Africa including Morocco, Algeria and Tunisia, all areas where it enjoys the cool, moist winters and hot, dry summers. The slightly glossy leaves are spiny, and the acorns ripen in autumn. These are foraged by pigs in Spain and Portugal, giving the resulting ham a particularly appreciated

The evergreen cork oak has glossy, spiny leaves and often twisted branches. In addition to its useful bark, it also provides acorns, which are foraged by pigs in Spain and Portugal. The resulting ham, called *jamón ibérico de bellota*, has a distinctive flavour.

flavour. But it is the unmistakable thick, rugged and deeply ridged bark for which this tree is most renowned. With this bulky covering of bark, *Quercus suber* is a pyrophyte, meaning it has adapted to tolerate fires. The bark contains high levels of a waxy substance called suberin, named after the specific epithet of the cork oak; this forms a protective barrier to the movement of water and solutions, providing one of the properties that make cork such a useful material.

The bark is first harvested when a tree is about twenty-five years old and has reached a specific size. This operation is carried out from May to August by trained, highly skilled harvesters by hand, using only a sharp axe of a traditional shape, with no machinery involved. Large, curved sections of bark are removed from the trunk without harming, killing or cutting down the tree in the process, making it a sustainable crop and renewable resource. Once the bark has been removed, it regenerates and can be stripped again every nine years, and generally only twelve times during a tree's lifetime. The exposed inner bark is a dark reddish colour, and a white number, one to nine, is painted on the trunks to indicate which year in the cycle the bark was last removed.

The first harvest, known as *desbóia*, yields 'virgin cork', which is too irregular in structure to be used for bottle stoppers and instead finds

ABOVE Harvesting the bark of cork oaks is traditionally done by hand, using only a sharp axe of a particular shape, as in this nineteenth-century illustration. The process is still the same today and usually begins when a tree is about twenty-five years old.

OPPOSITE Cork oaks are attractive trees with evergreen leaves, unlike many other species of oak, which are deciduous. They often occur in mixed woodlands, which provide valuable habitats for a wide range of wildlife.

BUILDING AND CREATING

many applications including for sound proofing, floor and wall tiles, pin boards, cricket ball cores and fishing rod handles. It is only from the third stripping that the cork has the best properties for wine and champagne bottle stoppers. Across Europe, the cork industry produces 300,000 tonnes of cork each year, employing around 30,000 people, with 15 per cent of the cork being made into bottle stoppers. So highly regarded are cork oaks that in Portugal, which produces 50 per cent of the world's cork, it is illegal to cut down a living tree, and special permission is needed to fell old and unproductive trees.

It is cork's impermeability to liquids and gases and its ability to be compressed and recover, as well as its lightness, that make it so successful for sealing wine bottles, but its insulating properties are also valued and exploited. From medieval monasteries to NASA's space programme, cork has long been used as insulation to protect against heat and cold. Cork taken from 225 cork trees in Portugal was used, with other materials, to insulate the space shuttle Columbia's external fuel tank.

Cork oak forests are also truly beautiful habitats. Trees are often grown with other species including the evergreen holm oak (*Quercus ilex*), stone pine (p. 110) and olive (p. 90) in a Mediterranean mixture of shrub land and pasture or *maquis*. These open woodland agroforestry systems, known as *montado* in Portugal and *dehesa* in Spain, contain a combination of open-grown trees, between 50 and 300 per hectare,

and aromatic understorey shrubs such as lavender, cistus, gorse and broom, which are normally associated with a Mediterranean garden rather than a wild environment. The natural stands of cork oaks, with their broad canopies, provide shade from the midday sun in summer and are used for animal grazing. They also support diverse ecosystems and are the favoured habitat for many species, including the vulnerable Iberian imperial eagle and the most endangered wild cat, the Iberian lynx.

Despite all cork's advantages, wine producers are finding other methods and materials to stopper bottles. So should we continue to use cork? The answer clearly is yes, for as long as there is a demand for this sustainable and renewable resource, there will always be a continued need for these natural orchards. If this need disappears, so will these trees, along with everything they support. In this way cork and conservation go hand in hand.

Cricket bat willow

Salix alba var. *caerulea*

There are over 300 species of willow found growing naturally in many parts of the world, ranging from tiny Arctic ones to large trees including the famous 'weeping' willow, *Salix babylonica*. Generally very fast-growing, willows' supple, whippy stems have been used for millennia to weave into baskets, fences, fish-traps and the frames of coracles and canoes. More recently they have been grown as a renewable and sustainable source of bio-fuel, and the timber is also used for charcoal and paper-making. But the wood of one willow has a very particular and special application, with an international reach.

The wood of the cricket bat willow ... is very tough and strong and does not splinter easily, a quality definitely needed when hitting a solid ball at high speed.

Despite the rules of cricket being some of the most complicated to understand for the uninitiated, in a nutshell the sport can be summed up in a brief sentence. The bowler delivers the ball from one end of the pitch to a batsman at the other, who, armed with a bat, strikes the ball far enough away to be able to make runs from one end of the pitch to the other. The wooden bat takes the shape of a blade with a cylindrical handle at the top; the total length of the bat can be no more than 96.5 centimetres (38 inches) and the blade itself can be no more than 108 mm (4¼ inches) wide. There is no specified weight, but bats are usually between 1.1 and 1.4 kilograms (2½ to 3 pounds). By far the most preferred wood to make bats is that of *Salix alba* var. *caerulea*, which is known therefore as the cricket bat willow, though sometimes called the blue willow.

This variety is thought to be a hybrid of the white willow, *Salix alba*, and the crack willow, *Salix fragilis*, but is most probably a variant of the white willow, which gets its name from the silvery appearance of its long, narrow leaves. The cricket bat willow makes a handsome

The cricket bat willow makes a handsome tree up to 30 metres (100 feet) tall with a pyramidal habit. It is cultivated for bat making, but also grows naturally in the eastern counties of England along water courses and in places where the water table is high.

tree to 30 metres (100 feet) tall with a pyramidal habit, more upright in form than the true white willow, and is found growing naturally in the eastern counties of England along water courses and in meadows where the water table is high. It is also cultivated today as a unique timber crop for the production of cricket bats. In moist but free-draining, fertile soil a tree can grow to a suitable size for harvesting in fifteen to twenty years. Most growers of these trees sell them still standing to specialist companies, who then carry out the felling and extraction themselves to prevent the trunks from splitting or being damaged. The world's oldest and largest company, established in 1894, is J. S. Wright & Sons, created by Jessie Samuel Wright after meeting a man called Montague Odd who was looking for willow trees in the area. Odd made bats for the great cricketer W. G. Grace for a guinea each, and his father, Amos, had perfected and patented the cricket bat from the non-standard form it was then to the familiar bat used today in international cricket.

The wood of the cricket bat willow has a low density, is light in weight, very tough and strong and does not splinter easily, a quality definitely needed when hitting a solid ball at high speed. Young trees are planted between November and March as four-year-old unrooted 'willow sets', which are produced from disease-free and registered coppiced stool beds of superior pedigree mature trees. The prepared sets are pushed into a hole made in the ground in an upright position, watered in and firmed. During the trees' short life, side shoots are rubbed off and any branches that develop are pruned off flush to the trunk before they become woody. This prevents knots developing in the wood that would make it unsuitable for the production of a bat. When felled, the

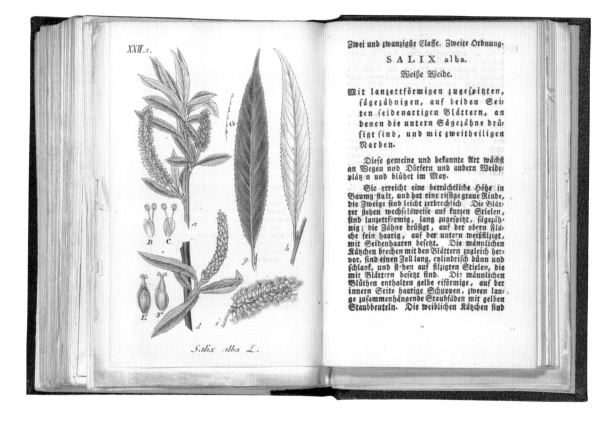

XXII.2.

Zwei und zwanzigste Classe. Zweite Ordnung.

SALIX alba.

Weiße Weide.

Mit lanzettförmigen zugespißten, sägezähnigen, auf beiden Seiten seidenartigen Blättern, an denen die untern Sägezähne drüsigt sind, und mit zweitheiligen Narben.

Diese gemeine und bekannte Art wächst an Wegen und Dörfern und andern Weideplätzn und blühet im May.

Sie erreicht eine beträchtliche Höhe in Baumgstalt, und hat eine rißige graue Rinde, die Zweige sind leicht zerbrechlich. Die Blätter stehen wechselsweise auf kurzen Stielen, sind lanzettförmig, lang zugespitzt, sägezähnig; die Zähne drüsigt, auf der obern Fläche fein haarig, auf der untern weißfilzigt, mit Seidenhaaren besetzt. Die männlichen Käßchen brechen mit den Blättern zugleich hervor, sind einen Zoll lang, colindrisch dünn und schlank, und stehen auf filzigten Stielen, die mit Blättern besetzt sind. Die männlichen Blüthen enthalten gelbe eiförmige, auf der innern Seite haarige Schuppen, zween lange zusammenhängende Staubfäden mit gelben Staubbeuteln. Die weiblichen Käßchen sind

Salix alba L.

trunk is cut into lengths called 'rolls', which are then split into 'clefts' before drying and being manufactured into a bat blade. As part of a sustainable regeneration programme, new sets are planted for continued supplies. Clefts are now exported to India and Pakistan, and *Salix alba* var. *caerulea* is grown in Australia for cricket-bat making.

Since 1924 in Britain, cricket bat willows have been under threat from what is known as 'watermark disease', caused by a bacterium (*Erwinia salicis*) which causes a discolouration and weakening of the wood, making it fracture easily and therefore unsuitable for the manufacture of cricket bats. Fortunately, the disease has been reduced to manageable levels, ensuring the continued supply of cricket bats so that players and spectators all round the world can still enjoy the sound of leather on willow, which has inspired poets and authors from John Betjeman to A. A. Milne.

Eychbaum.

CXXIX.

Quercus Robur L.

BUILDING AND CREATING

English oak

Quercus robur and *Quercus petraea*

Of the around 600 species of oak that grow in the temperate northern hemisphere, two have a native range across most of Europe and into the Caucasus: *Quercus robur*, the English, European or pedunculate oak, and *Quercus petraea*, the sessile oak. The two species are similar in overall appearance, but can be distinguished by key botanical features reflected in their common names. *Quercus robur* has short petioles, or leaf stalks, and long stalks – peduncles – holding the acorn cup, so is known as pedunculate, while *Q. petraea* has long-stalked leaf petioles and sessile, short-stalked acorn cups borne directly on the twigs, and is therefore known as the sessile oak. *Quercus robur* is an iconic tree that almost everyone, of every age, can easily recognize from its typical round-lobed leaves and familiar acorn fruits. The species name *robur* reflects the strength and durability of the tree's hard timber.

Both these species of oak are impressively long-lived, large, deciduous trees growing to over 30 metres (100 feet) high, with wide-spreading crowns. It is thought that there are around 121 million oak trees in the woodlands of the United Kingdom, and they are the most frequently found tree species among open-grown, individual specimens, with almost 1 million trees in London alone. These oaks can live for around a thousand years if pollarded (a system of pruning the branches), but an average lifespan is thought to be around 250 years. An oak tree that survives to reach 400 years old begins to decline in vigour, naturally reducing in size and height through the ageing process and acquiring a squat shape and a wide, hollow trunk. England in particular is home to a large number of these 'ancient oaks', as they are known. It is estimated that there are at least 3,300 ancient native oak trees in England – more than the total found growing in all other European countries put together.

Quercus robur, the European or English oak, with its familiar leaves and acorns, is one of around 600 species of oak that grow naturally in the temperate northern hemisphere.

These ancient trees have been preserved in England for a number of historical reasons, apart from the ideal growing conditions. In the eleventh century William the Conqueror established a system of forest law, creating Royal Forests to conserve the habitat for game and wild animals, which could be hunted only by the king, while the nobility were granted 'chases' and deer parks. It was forbidden to fell any trees in these forest reserves and in this way they could be seen as an early form of nature conservation. Later historical developments also favoured the continued preservation of the oak trees of England, including private ownership of parks, the availability of timber supplies from overseas and the absence of destructive wars in the landscape. Finally, modern forestry practices were introduced too late to lead to the felling of these old, hollow – and therefore useless – timber trees. They may be no good for timber, but they are the basis of an important ecosystem and provide habitats for a vast range of flora and fauna, supporting over 2,000 species of fungi, bryophytes (mosses, liverworts etc), lichens, insects, birds and animals. Oaks are said to be associated with more than 300 obligate species of wildlife – specialized organisms that live only on this tree – more than any other native British species of tree.

For almost two centuries the wood of this oak has been used by coopers to make the barrels needed for the production of Scottish whisky.

Many individual ancient oaks have evocative or descriptive names and seem to possess a character of their own. One of the largest, oldest, at around 900 years, and most famous is the 'Major Oak' in Sherwood Forest, with a girth of 10.7 metres (35 feet). Robin Hood and his Merry Men are said to have sheltered from the Sheriff of Nottingham under its broad canopy, which spreads over 28 metres (92 feet) wide and is now supported with metal props to prevent it collapsing. And in the middle of a farmer's field near Bourne in Lincolnshire is the 'Bowthorpe Oak', possibly the largest oak tree, with a massive girth of 12.8 metres (42 feet). Over the years it has become completely hollowed out by fungal decay, and as far back as 1768 it was used as a pigeon house and then an outdoor eating place complete with a door and seating for up to twenty diners, which have long since disappeared.

Because of the strength and durability of its wood, the oak has helped shape British history through its use in house-building, furniture-making and ship-building before steel. In 1512, the *Mary Rose*, a carrack-type warship of Henry VIII's Tudor navy, was built using timber from around 600 large oak trees; when growing these would have covered roughly 16 hectares (40 acres) of woodland. Over two hundred years later, in 1765, more than 5,000 oak trees provided

Ancient oak trees often have fantastically twisted and gnarled forms and hollow trunks. It is estimated that there are more of these trees in England than the total found growing in all other European countries. Many individual ancient oaks have evocative or descriptive names that tell something of their history and associations.

BUILDING AND CREATING

QUERCUS racemosa.   CHÊNE à grappes

P. Bessa pinx.   Gabriel sculp.

the timber for HMS *Victory*, Nelson's flagship at the Battle of Trafalgar in 1805. Oak was also a major component in the construction of high-status buildings: over 660 tons of oak timber were used to build the magnificent hammer-beam roof of Westminster Hall in London in 1393. While in central France the great Forêt de Tronçais still contains sessile oaks planted in 1670 by order of the Minister of Finance of Louis XIV, Jean-Baptiste Colbert, to provide timber for the French navy far in the future.

Of the many insects that live on the oak tree, one tiny wasp (*Andricus kollari*) has large implications for human civilization. It is the origin of the oak marble gall on the leaf buds, a spherical abnormal growth caused by the irritation of the insect. This gall contains large quantities of tannic acid, one of the main ingredients for the production of 'iron gall ink', used in writing for over 1,800 years. Many important works such as the Dead Sea Scrolls and the Magna Carta were written with this ink, and Newton penned his theories and Mozart his music using oak gall ink.

For almost two centuries the wood of this oak has been used by coopers to make the barrels needed for the production of Scottish whisky, which has to be matured in an oak cask for at least three years before it can legally be called whisky. The toughness of the wood means that the barrel staves can be bent by heat without splitting, while its tight grain prevents the liquid leaking, but is porous enough to allow the movement of oxygen in and out of the cask and oils or 'vanillins' to be drawn out of the wood, influencing the whisky's flavour. For similar reasons, oak barrels are also used in the ageing of wine around the world, imparting desirable flavours and character to the end result.

European oaks are unfortunately under threat from several exotic pests, including the oak processionary moth (*Thaumetopoea processionea*). The moth's caterpillar can defoliate mature oaks, weakening them and making them vulnerable to other harmful organisms. Diseases such as acute oak decline, a bacterial infection associated with a bark beetle (*Agrilus biguttatus*) can lead to the death of a mature tree within four to five years. It would be a great travesty if we allowed our oaks to succumb to such threats. Their loss would have serious impacts on our well-being, economy and environment, as well as on all the species that the trees support. We must do all we can to prevent the decline in the health of our oaks, so that future generations can continue to enjoy these magnificent trees.

Foxglove tree

Paulownia tomentosa

The genus *Paulownia* has a distinguished pedigree – it was named by the eminent nineteenth-century German botanist Philipp Franz von Siebold in honour of Her Imperial Highness Grand Duchess Anna Pavlovna of Russia, the eighth child of Tsar Paul I of Russia and the Empress Maria Feodorovna. There are between seven and ten species of *Paulownia*, mostly native to western and central China, with two species native to the Taiwan region. The rare Taiwanese species *Paulownia kawakamii* is now classed as critically endangered, with fewer than 100 specimens occurring in the wild. However, the more common Chinese species found in cultivation around the world is *Paulownia tomentosa*. Like many trees, *Paulownia* has several common names that hint at its characteristics and the historical stories associated with it. They include empress or emperor tree, sapphire dragon tree, the princess tree and the foxglove tree.

> *Of all the ornamental temperate trees grown in gardens, this is a botanical oddity that stands out from the crowd, particularly in early spring when it flowers.*

Of all the ornamental temperate trees grown in gardens, this is a botanical oddity that stands out from the crowd, particularly in early spring when it flowers. Its large, lavender-purple foxglove-like flowers – hence one of its names – are held in upright panicles, up to 40 centimetres (16 inches) tall, on the tips of the branches before any leaves appear. The swollen brown flower buds are produced in the previous summer, so the flowering display is always more impressive after a mild winter, as the buds are susceptible to frost damage.

Large, soft, felty, ovate leaves are produced after the flowers and are covered in hairs, giving this tree its specific epithet *tomentosa*, Latin for 'woolly' or 'covered in hairs'. In the garden, horticulturists often take advantage of these large leaves by coppicing the tree's main branches

Paulownia kawakamii, also known as the sapphire dragon tree, is one of seven to ten species of *Paulownia*, many of which are from China. This rare species is now classed as critically endangered, though in cultivation it can be very fast growing.

FOXGLOVE TREE

down to ground level to encourage vigorous new shoots with huge, dinner-plate-size leaves up to 80 centimetres (32 inches) across.

If left to grow as a tree, paulownia can reach up to 8 to 12 metres (26–40 feet) tall and is quick to mature. In Japan this fast-growing tree is known as *kiri*, and its high-quality, light and fine-grained timber is used for particular purposes. It was once the practice for elite aristocratic families to mark the birth of a baby girl by planting one or several paulownia trees so that they would grow and mature in time to be cut down to make her dowry chest and other furniture when she married. In Kyoto, Nara and Osaka it is still customary for the bride's father to present his daughter when she moves into her new marital home with a chest or kimono dresser carved from paulownia wood, for storing her delicate silk garments. The valuable white wood is also typically used to make several Asian musical instruments including the Japanese *koto* and the Korean *gayageum* zither. Paulownia also has a long history of use in China, where, as in Japan, it is associated with longevity. One Chinese legend tells how the Phoenix perched only on paulownia trees as it flew from the south sea to the north sea, so the tree is often planted near homes to bring good luck.

The numerous cylindrical woody fruits produced after flowering contain thousands of light, soft, winged seeds. Before the advent of polystyrene, these were used as packing material in cases of Chinese porcelain for export. The cases would often break open in transit, leaking out the seeds which were then distributed and spread by

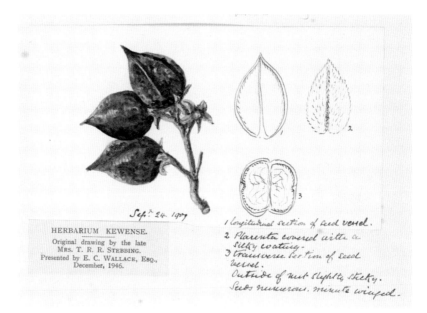

BUILDING AND CREATING

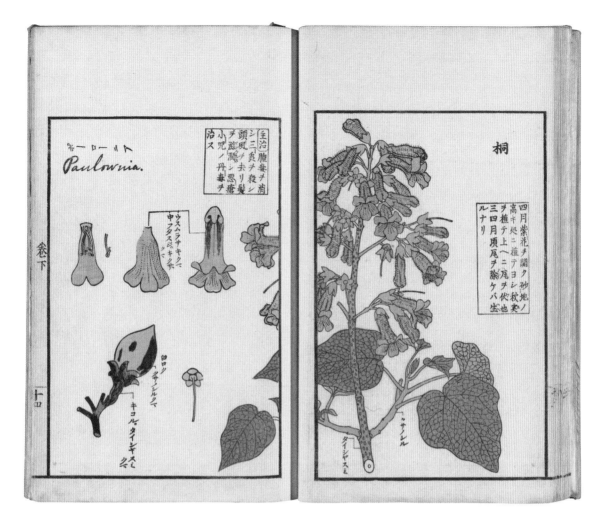

wind along the major transport routes, where they would germinate. The resulting trees would grow quickly and out-compete the natural vegetation, becoming a weed problem. Today the seeds even germinate in small cracks in pavements and walls, soon smothering other plants with their large leaves. In many parts of the world where the climate suits the growth of this tree, particularly Japan and the eastern United States, paulownia is classed as an invasive species.

From as early as the twelfth century the paulownia seal was the private symbol of the Japanese imperial family, before the chrysanthemum seal. Following the Meiji restoration in 1868, the seal was used as the emblem of the Japanese government. Today the stylized paulownia government seal is used by the Prime Minister of Japan on all official documents.

Rotnuß.
Haßlnuß

CCXXIIII.

Nuces Avelanæ rubræ
corylus Avellana l.

Hazel

Corylus avellana

In the bleakness of late winter or early spring, before any fresh green leaves reappear, deciduous woodlands across Europe and western Asia, from the British Isles to the Caucasus, Turkey and northwestern Iran, are brought alive with the appearance of dancing pale yellow 'lambs' tails', the male flowering catkins of hazel. The common hazel, *Corylus avellana*, is a large shrub or small deciduous tree with multiple trunks and basal branches growing from just above ground level. The genus name *Corylus* is derived from the Greek word *korys* meaning 'helmet', referring to the outer ruff or husk enclosing the nut, while the specific epithet *avellana* comes from the town of Avella in the Campania region of Italy. This was selected by the great naturalist and taxonomist Carl Linnaeus from Leonhart Fuchs's 1542 herbal *De historia stirpium* ('Notable commentaries on the history of plants'), in which the hazel was described as *Avellana nux* or the 'nut of Avella'.

> *Grimm's Fairy Tales claims that a 'green hazel branch has been the safest protection against adders, snakes and everything else which creeps on the earth'.*

When cultivated for its wood, hazel is grown and managed in woodland parcels known as 'coupes'. The hazel trees are sustainably cut back in rotation every five to ten years to promote the rapid growth from the base of shoots or rods. These provide a renewable raw material with a multitude of traditional uses. The rods are supple and easy to split with hand-tools, and from the earliest times have been turned into woven wattle panels for houses and fences, spars for thatched roofs, the frames of coracle boats, and hedge stakes and bindings for strengthening field hedges. The tips and finer ends of the stems are bundled into faggots for fuel in kilns and the twigs are used as sticks to support peas and beans; nothing is wasted. Smaller diameter rods,

Carl Linnaeus took the species name for the common hazel, *avellana*, from the 1542 herbal of German botanist Leonhart Fuchs, which included this illustration. In the text, the hazel was described as the nut of Avella, an Italian town.

typically 1.25 metres (50 inches) long that have been drawn up straight to the light are harvested as shanks for the production of walking sticks. This finds an echo perhaps in the medieval practice of placing staffs or wands of hazel, ash or willow in burials to act as a protective charm.

Pollen from the male catkins is carried by the wind to the hazel's female flowers, which are bright red but very small and inconspicuous. The resulting edible hazelnuts or cobnuts ripen in late summer. Roughly round to oblong-shaped, the nuts form an important commercial crop and are rich in protein, vitamin E and unsaturated fats. There are many cultivated forms of hazel, bred for their nut size and early or late harvest. Filbert nuts, which are larger and more elongated than the cobnut, are the fruit of a different species, *Corylus maxima*, and are named after St Philibert of France, whose feast day is 20 August, the peak harvest time of hazelnuts. The top commercial producer of hazelnuts today, accounting for 75 per cent of the world's production, is Turkey, where around 600,000 tonnes of nuts are harvested each year. These are used for making a wealth of confectionery products, from hazelnut marzipan to hazelnut chocolate spread.

A different species of hazel, *Corylus cornuta* var. *californica*, is native to the Pacific Northwest of America, and was a food source for indigenous peoples. The European hazel was introduced to North America by settlers from France and England, and is an important crop in the state of Oregon, which is home to 99 per cent of America's commercial hazelnut growing. Trials are now also being carried out to grow *Corylus avellana* in China.

Hazel coppice woodland is biodiverse and an important source of food for a number of animals and insects. It is also a key habitat for several birds and mammals, especially the enigmatic nocturnal hazel dormouse, *Muscardinus avellanarius*. As its name suggests, hazelnuts are the dormouse's principal food for fattening up before hibernation, but the continuing loss of ancient woodland and a decline in traditional management practices mean that populations of this mammal have decreased by a third over the past two decades.

Like many trees, hazel appears in mythology and folklore. 'The Hazel Branch' from Grimm's Fairy Tales claims that 'a green hazel branch has been the safest protection against adders, snakes and everything else which creeps on the earth'.

HAZEL

Paper birch

Betula papyrifera

Also known as the canoe birch or white birch, the paper birch grows naturally in sandy, moist soils by rivers in full sun across North America, from the Atlantic to the Pacific coast and from the states of Colorado and Virginia to Alaska. It is most common in the northern states and all the provinces and territories of Canada. A true pioneer tree, it is one of the first woody plants to re-colonize areas where forests have been cleared or destroyed by fire, and grows faster than any other birch. In an average year, approximately 1 million seeds per acre are produced, but in a bumper year this figure can rise to 35 million. The seeds are extremely light and are blown by the wind long distances to new, open areas, where the trees quickly establish and grow before other species arrive.

Numerous species of birch are found around the globe in the northern hemisphere. All are elegant and graceful trees, having a crown with a delicate, lacy habit and leaves that turn a deep golden yellow colour in autumn. In its natural habitat, the paper birch makes a medium-sized, deciduous tree to about 30 metres (almost 100 feet) tall, often with a single trunk, unless it is browsed by moose, as the fresh young shoots are a key part of the animal's diet. In a garden setting this birch is usually grown as a multi-stemmed tree in order to show off its main attribute more effectively – its attractive, white mature bark. The first person to describe the tree, Humphry Marshall, a cousin of the American naturalist John Bartram, wrote that it has 'a very smooth white bark'. This description appeared in his *Arbustrum Americanum, The American Grove*, an alphabetical catalogue of trees and shrubs native to North America printed in Philadelphia in 1785, one of the earliest books published in America.

The timber ... has multiple uses, including for furniture, flooring, ice lolly sticks, tooth picks, veneers and plywood sheets.

The paper birch is a true pioneer tree, which means it is one of the first woody plants to re-colonize land where forests have been cleared or devastated by fire. It can also spring up in cracks and crevices in rocky ground, sometimes resulting in a contorted shape.

The specific epiphet *papyrifera* is derived from the Greek word for papyrus paper and the Latin *ferre*, which means to bear or carry, and so translates as 'paper-bearing'. This very aptly refers to the snow-white bark, which appears as the tree starts to mature and peels off the trunk in thin, paper-like layers and has in fact been used as a form of paper. The young bark begins its life a reddish-brown colour, which can make identification of this tree in the wild confusing. A very high oil content means the paper birch's bark is both waterproof and very weather resistant, often outliving the wood of the tree itself when it is felled. It is the perfect fire starter, igniting even when wet.

As the bark is so durable and flexible, as well as waterproof, it has been of great importance to Native Americans and First Nations Peoples, who use it in many different ways. Once they have carefully peeled bark from the trees to avoid harming them, the indigenous Anishinaabe peoples of the Upper Great Lakes region surrounding Lake Superior sew sections together to make a *wiigwaasi-makak*, a birch-bark container or box. A cord called *watap*, made from the stripped roots of conifers growing alongside the birch, is used to stitch the pieces together. The containers are used mostly to collect and store food and are still made today to sell to visitors. Paper birch bark could also be applied as a durable waterproof layer to sod-roofed houses. Many indigenous groups such as the Wababaki people of Maine used it to make wigwams, everyday utensils and lightweight canoes that could be easily carried between river systems, from which the tree gets another of its common names, the canoe birch.

In Alaska the sap from the paper birch, known as birch water, is tapped and collected in the forests from living trees during early spring, before the leaves appear. It contains about 1 to 1.5 per cent sugar, and is reduced down to make 'birch syrup', similar to maple syrup. Between 100 and 150 litres (22 to 35 gallons) of sap are required to make 1 litre of syrup.

The timber of *Betula papyrifera* is a moderately heavy, white wood and also has a wide range of uses, including for furniture, flooring, ice lolly sticks, tooth picks, veneers and plywood sheets. Other items traditionally made from the wood of the paper birch are spears, bows,

ABOVE AND OPPOSITE The species name of paper birch, *papyrifera*, translates as 'paper-bearing', and the distinctive white peeling bark can in fact be used as paper, as seen in the books opposite. It can also be used to make birchbark containers, wigwams, a variety of utensils and even lightweight canoes.

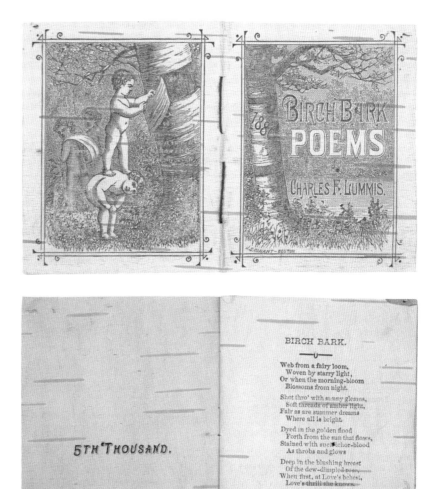

arrows, snowshoes and sledges, among many other utensils and objects. In addition, the dried wood, if seasoned properly, makes excellent, high-yielding firewood. This valuable tree also has many traditional medicinal applications, for instance in the treatment of gout, colds and coughs, pulmonary diseases and even rheumatism. Moreover, it is an effective laxative, is considered helpful in easing burns and healing wounds, and is now being investigated for its potential in the treatment of cancer.

Sitka spruce

Picea sitchensis

===

In 1827 on Sitka Island, the eighth largest island in Alaska and known today as Baranof Island, the German botanist and naturalist Karl Heinrich Mertens collected specimens of a tree. Mertens is himself recognized in the name of the mountain hemlock, *Tsuga mertensiana,* which he also collected specimens of, but this particular tree was named after the island, *Picea sitchensis,* the Sitka spruce, now regarded as a foremost timber tree. Sitka spruce was actually first described botanically by Archibald Menzies in 1792 on the shores of Puget Sound, off the coast of Washington State. It was later introduced into cultivation in the British Isles named *Pinus menziesii* from seed collected by David Douglas in 1831 on his third and final expedition to North America, but was later given the name *Picea sitchensis* by August Gustav Heinrich von Bongard and Élie-Abel Carrière in 1832.

The roots were gathered by Tlingit and Haida to make waterproof baskets and hats, and provided the material for ropes and fishing twine.

Sitka spruce has an incredible natural range, covering 2,900 kilometres (1,800 miles) north to south along the Pacific Northwest of the USA and Canada. The southernmost tree is found in Mendocino County, California, and the northernmost in Prince William Sound in Alaska, where it is the state tree. The coastal belt in which it grows is around 400 kilometres (250 miles) wide east to west. This tree is also a near record breaker, being the fifth largest and third tallest conifer in the world, and the largest spruce. It has an imposing conical shape, with a pointed crown and straight trunk that can measure just under 100 metres (330 feet) tall and 5 metres (16 feet) in diameter at the base. The branches are graceful, sweeping outwards, and bear stiff, prickly needles which are a bluish-green colour. Reddish-brown seed cones hang from the drooping ends of the branches. Sitka spruce are

The Sitka spruce can grow up to 100 metres (33 feet) tall, making it the third tallest conifer in the world. It is often now grown in ranks in plantations for its valuable timber, but a specimen tree has a conical shape with sweeping branches borne on a straight trunk.

ABIES MENZIESII

typically found in cool and humid upland locations, places where not many other trees can survive or grow well. They require abundant moisture, whether in the form of rainfall or the coastal fogs of their natural habitat.

The fresh young tips of the foliage are a source of vitamin C and were used by indigenous peoples to make spruce beer as a cure for scurvy during the winter months, when fresh fruits were not available. The roots were gathered and woven by peoples of the Northwest Coast, including Tlingit and Haida, to make waterproof baskets and hats, and also provided the material for ropes and fishing twine, while the trees' resin was used for caulking canoes or as a glue.

Sitka spruce is now extensively planted as a timber tree in plantations in northwest Europe, particularly Britain, Norway, Denmark and Iceland. As a forestry crop, this tree's ability to grow in harsh environmental conditions and poor, thin soils is one great advantage. Another is its fast growth rate, which means it can yield high volumes of usable timber in a relatively short time. Trees reach their maximum timber potential in as little as 40 to 60 years, compared with an oak which takes more than 150 years.

Sitka's great qualities as a timber are its high strength and stiffness relative to its density and weight. The wood is also regular in grain and knot-free, making it an excellent conductor of sound, which is why it is used in acoustic musical instruments, for instance for the sound-boards of violins, guitars, harps and pianos. The energy of the vibrating strings is transferred through the wood, which resonates beautifully. For such uses 'old growth' wood from trees at least 250 years old is pre-ferred. The combination of strength and lightness also made Sitka an

important material in aircraft building. The Wright brothers used it in their Flyer, and many later air-craft, including the British de Havilland Mosquito, incorporated Sitka spruce in the wing spars. For the same reasons, boat builders regard it as second to none for the masts and spars of sailing yachts. The timber is also used in the construction industry, and early thinnings from forestry plantations make strong, smooth paper because of the white-coloured wood and long cellulose fibres.

One unusual Sitka with golden foliage that appeared to glow in the sun, named 'K'iid K'yaas' or 'ancient tree', and also known as 'The Golden Spruce', grew on the banks of the Yakoun River on the Haida Gwaii archipelago in Canada. It was

sacred to the Haida people, but was illegally cut down in 1997 in protest at the logging industry. The world's largest tree today grows on the Western Olympic Peninsula in Washington, near Lake Quinault in 'The Valley of the Rain Forest Giants'. A true champion, it is estimated to be over 1,000 years old and stands at 58.2 metres (191 feet) tall with a trunk diameter of 5.8 metres (18 feet 9 inches). It is an understatement to say that this is an impressive tree, and who knows what the trees planted in more recent times in forestry settings in Europe might look like in a thousand years' time?

Yew

Taxus baccata

═══

A tree with often rather dark and gloomy associations, *Taxus baccata*, the yew, is a medium-sized evergreen that can be found growing naturally in western, central and southern Europe, through to Iran and the Caucasus. There are many theories as to where the name for the genus, *Taxus*, originates, but it is thought that it may be derived from the ancient Greek word *toxon*, 'bow'. The Roman authors Virgil and Pliny certainly mention the use of yew wood for bows and poisoned arrows in their works. The specific epithet *baccata* is Latin, meaning 'having fleshy berries', despite the fruit of the yew not being a true berry but a seed surrounded by a bright red aril or coating. This juicy, sticky, sweet-tasting flesh is the only part of the tree that is not poisonous, as everything else, including the seeds and leaves, is toxic to humans and animals because of taxine alkaloids. Between 50 and 100 grams (1¾ to 3½ ounces) of yew leaves is considered to be fatal to an adult.

The Druids regarded the yew as a symbol of everlasting life and planted trees close to their temples.

Yews are slow-growing, venerable trees which generally live for centuries, with an average age of about 500 years. They can reach up to 20 metres (66 feet) tall, with massive trunks and dense, spreading branches. Ancient trees are often seen standing alone in churchyards, and these can grow into much larger and more impressive specimens than trees found in natural woodland settings as they have more light and space to develop. They are also thought to be much older, with some estimated to be up to 2,000 years old or even more, though this is a question surrounded by much speculation and dispute. Their age may be overestimated because of their size and appearance of great antiquity, and often cannot be measured accurately by counting the annual growth rings as yews tend to hollow out, losing the internal heartwood.

Although the yew is classed as a conifer, it does not produce cones. The seeds are instead encased in red, sticky flesh, known as an aril. This is the only part of the yew tree that is not poisonous. The dark green needle-like leaves and the seeds inside the aril are toxic.

Local records, including old maps, engravings and paintings, can be good sources of evidence, providing a snapshot in time of when a particular tree grew in that place.

Two of the oldest trees in Britain are churchyard yews. The Fortingall Yew in Perthshire, Scotland, is estimated to be around 2,000 years old, while the Defynnog Yew in Powys, Wales, is thought to be around 2,500 years old, though both are claimed to be as much as 5,000 years old, depending on which historical facts and accounts are believed, making the yew one of the oldest plants in Europe. Another celebrated yew is found in the churchyard of St George's Church in Crowhurst, Surrey. Today it has a girth of 10 metres (33 feet), and historical measurements dating back to the seventeenth century reveal that this has only increased by 65 centimetres (26 inches) in 369 years. It is so large that in the eighteenth century a hinged wooden door was built into its base to allow access into the hollow centre. In 1820 local villagers found a cannon ball embedded in the trunk, which is thought to have been there since the English Civil War, perhaps dating to 1643.

This link between yews, churches and graveyards is a very ancient one, perhaps reaching back into pre-Christian times, though the numerous suggested explanations for the tradition are hard to substantiate. The yew has funerary or underworld associations in many

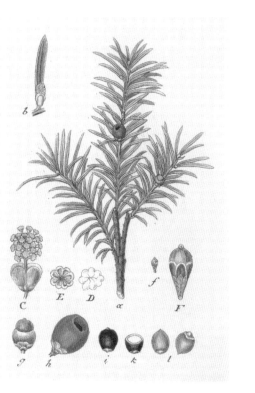

European cultures. In ancient Greek mythology it was sacred to Hecate, goddess of magic, witchcraft and the night. Yggdrasil, the tree of life in Norse myth, with its roots in the underworld and branches in the heavens, is usually said to be an ash tree, but others think it is a yew. The Druids regarded the yew as a symbol of everlasting life, and planted trees close to their temples – although the yews were dark and cold, they were also glorious and evergreen, and long-lived, which reinforced the Druids' belief in death and reincarnation. When Pope Gregory I instructed St Augustine and his missionaries in the sixth century to convert pagan Britain to Christianity, churches were built close to the pagan sites of worship, with their yew trees, in the hope of encouraging the pagans into the warm, bright, airy churches. In the churchyard of the sixth-century St Brynach's Church in Nevern, Pembrokeshire, Wales, an ancient yew, known as the Bleeding Nevern Yew, is thought to be over 700 years old. For as long as anyone can remember, it has exuded a blood red sap from the trunk. There is no explanation for this botanical oddity, except the interpretation that the tree is bleeding in sympathy with Jesus as he was crucified on the cross.

The wood of yew is a rich orange-brown colour, and is among the hardest of the conifers (although it does not bear its seeds in a cone, yew is classified as a conifer). Our early ancestors clearly valued it – one of the oldest surviving wooden artifacts in the world is a yew spear head found in 1911 at Clacton-on-Sea, Essex, eastern England, which is estimated to be over 400,000 years old. And archaeologists at Greystones in County Wicklow in Ireland recovered a set of six carefully worked wooden pipes fashioned from yew, lying side by side in descending order of size and dating to around 2000 BC, which represent the world's oldest surviving wooden musical instrument.

Yew wood also has a remarkable elasticity and tensile strength, making it ideal for the English and Welsh longbow. At 1.8 metres

Single yew trees are often found in churchyards and it has been suggested that they were planted there to provide timber to make longbows, though they would never have supplied sufficient wood for this purpose. They do, however, frequently grow to a large size and a great age, as here at Buxted, Sussex.

BUILDING AND CREATING

(6 feet) tall, the longbow was one of the most effective weapons in medieval warfare, particularly in the battles of Crécy in 1346 and most famously the battle of Agincourt in 1415 during the Hundred Years War between the English and the French. A typical bow stave uses the natural lamination of the heartwood and sapwood to generate the power and velocity to shoot an arrow. The red heartwood, which resists compression, forms the inside or 'belly' of the bow, while the outer white sapwood provides the tension. With the increased demand for yew wood for longbows in England, stocks were depleted. There was a serious shortage of bow staves until Edward IV created the Statute of Westminster in 1472, which required every ship arriving at an English port to pay a tax of four bow staves for each tonne of cargo. In 1982, 130 longbows in excellent condition were recovered from Henry VIII's ship the *Mary Rose*, which sank off Portsmouth in the Solent in 1545. The bows had a draw strength of 100 to 185 pounds and a trained archer or bowman using one of these yew longbows could shoot between ten and twelve arrows a minute. The force was such that the arrows could wound the enemy at 365 metres (400 yards), kill at 180 metres (200 yards) and penetrate armour at 90 metres (100 yards).

Many believe that the yews in churchyards were planted there to provide wood for longbows, but it would take at least one hundred years for them to grow to the minimum size required, and the number of trees found in graveyards could not provide the amount of timber needed. In fact, yew wood was imported from the higher mountains of Europe as this had a tighter grain and was far more suitable and sought after for the longbow.

Yew has always been an important source of material for medical use. Most significantly, in 1967 Monroe Wall and Mansukh Wani of the Research Triangle Park in North Carolina discovered that certain compounds found in the bark of the Pacific yew, *Taxus brevifolia*, had anti-cancer properties. Unfortunately, the process of stripping the bark from the tree's branches kills it, and wild populations of the Pacific yew were under serious threat. It was clear that an alternative, sustainable source was needed and fortunately it was later discovered that the drug could be synthesized from the leaf clippings of the English yew, which is widely used in landscapes and gardens as a formal hedging plant and for topiary. Today the drug known as Paclitaxel, derived from taxol, is used successfully in chemotherapy for various forms of cancer. The myriad stories, myths and mystical and actual properties of yew all add to the intrigue and wonder of this tree.

Sweet chestnut

Castanea sativa

The sweet chestnut is a generous tree, often producing an abundance of fruits. In a timeless tradition redolent of cold winter days, sweet chestnuts are still roasted in their tough outer shell on an open fire or brazier, giving off an unmistakable aroma. The nuts can also be made into candied treats – the celebrated 'marrons glacés'. *Castanea sativa*, the sweet chestnut, is also known as Spanish chestnut or 'marron', which is French for chestnut, and is not to be confused with the unrelated horse chestnut, *Aesculus hippocastanum*. It is often said that the botanical name *Castanea* comes from the town of Kastania in Thessaly, Greece, where great numbers of trees were grown for their edible nuts, though it may be the other way round. The specific epithet *sativa* is the Latin botanical adjective meaning 'cultivated', used to describe domesticated plants that are grown as a crop.

Sweet chestnuts are long-lived trees when growing naturally in their preferred climate and soil, and can live for up to 2,000 years.

Native from the Balkan Peninsula to northern Iran and now found growing widely in southern, central and western Europe and into North Africa, the sweet chestnut makes a large, characterful, drought-tolerant tree up to 35 metres (115 feet) high. Trees often have huge, broad trunks up to 9 metres (30 feet) in girth, with deeply fissured, spiralling bark. The leaves are long and pointed, with serrated edges, and the comparatively rather insignificant yellow flowers are held in spikes. The fruits, however, are very distinctive, housed in a green, prickly case or 'bur' that protects them from predators before they are ripe. This then naturally splits open to expose the shiny, reddish-brown nuts inside – two to three, occasionally four.

In the eighteenth and nineteenth centuries the sweet chestnut was popular as an ornamental tree in the parklands of stately homes and large gardens, including the Royal Botanic Gardens, Kew, where some of the oldest trees in the arboretum are sweet chestnuts.

In southern Europe chestnuts, which are low in fat and calories, were highly valued as a food source and traditionally ground into flour.

The Roman legions marched partly on a diet of chestnut porridge. Numerous named varieties such as 'Marron de Lyon', 'Marigoule' and 'Bouche de Bétizac' have been bred and are today cultivated in orchards for commercial production. Chestnuts are still popular – the town of Collobrières in southern France is renowned for its annual festival, where the nuts are made into every kind of delicacy.

Sweet chestnuts are long-lived trees when growing naturally in their preferred climate and soil, and can live for up to 2,000 years. The oldest and largest known specimen is the 'Hundred-Horse Chestnut' or 'Castagno dei Cento Cavalli', which grows on the eastern slope of Europe's tallest volcano, Mount Etna in Sicily. Its exact age is not known, as after the girth was measured at 57.9 metres (190 feet) in 1780 the trunk has split into three large, separate sections; however, there are claims that it is between 2,000 and 4,000 years old. The name originates from a legend that tells how the Queen of Aragon and her one hundred knights took shelter under the tree during a severe thunderstorm.

In addition to its tasty nuts, the sweet chestnut is also highly valued for its wood. Trees are grown in woodlands where they are coppiced every twelve to thirty years, producing a sustainable crop of timber. The tannins in the wood mean that it is long lasting and durable when in contact with soil and therefore ideal for using outdoors. The timber is light in colour, hard, very strong and splits easily with wedges, making it perfect for fencing posts and rails. It is also used for cladding, and in parts of France church steeples are often covered in chestnut shingles. In southern Europe sweet chestnut is favoured by coopers to make barrels for ageing balsamic vinegar.

There are many other species of *Castanea*, including the American chestnut tree, *Castanea dentata*, which was formerly a major component of forests in the eastern United States. This chestnut was particularly important in Appalachia, where the nuts were foraged for food, and the timber and tannins from the bark were also utilized. In the early twentieth century a fungal disease originating in eastern Asia called chestnut blight (*Cryphonectria parasitica*) was discovered in New York and

within forty to fifty years had spread across the native range of the American chestnut from Georgia in the south to Maine in the north, destroying an estimated four billion American chestnut trees. The disease is now also widespread in Europe, and like the North American chestnut, the European chestnut is susceptible. Fortunately, Japanese chestnut (*Castanea crenata*) and Chinese chestnut (*C. mollissima*) are more resistant to blight and also yield edible nuts. They are sometimes used as attractive street trees, though they have one disadvantage – the prickly fruits are painful if trodden on in bare feet.

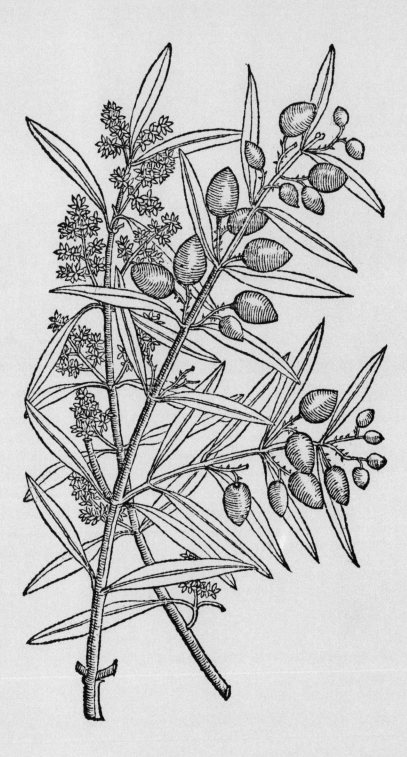

Olive, *Olea europaea*

FEASTING AND CELEBRATING

We have been enjoying the bounty provided by trees in our feasts and celebrations since time immemorial. From the jewel-like fruits of the pomegranate to the seeds of the cones of the stone pine, the stem of the sago palm and the aromatic bark of the cinnamon tree, humans have been very creative in seeking out, adapting, cultivating and harvesting crops from trees. Many have been elevated from staples to luxuries and are firmly embedded within the customs and heritage of numerous cultures.

The stories of how we discovered that various tree species offered such delights form a fascinating treasury of botanical history, myth and tradition. Our love of some, such as cacao and olive, can be traced back thousands of years and these trees are closely identified with particular civilizations. Our desire for others, such as the spices nutmeg and mace, changed the fortunes and lives of thousands of people, and were so valuable that obtaining them involved intrepid exploration, the creation of new trade routes, and even war, slavery and death.

Of course trees do not provide fruits, nuts or bark simply to fulfil our needs and desires – they all have a purpose for the trees themselves. The fruits exist purely as a method to disperse seeds from the parent tree to create the next generation, and take the form of attractive packages to lure creatures to eat them and store them, thereby helping the seeds to move further away from the tree than would otherwise be possible. Birds as well as bats, possums, rodents, monkeys and other mammals are great dispersers, eating the fruits and seeds and then depositing them elsewhere, with natural fertilizer to encourage germination and growth.

We have become the ultimate consumers of fruits and are adept farmers, spreading species by our own means. Plantations of tree crops fulfil our prodigious appetite for the fruits and seeds of trees. The coconut is now thought to be grown in over eighty countries around the world, and is no longer found naturally in the wild. A native of China, the persimmon has been taken to many new lands, including Korea, Japan, America and Europe for its delicious fruit. On the other hand, a few species have resisted domestication by people, with the Brazil nut being the prime example. Its pollination ecology is so complex and interconnected with its natural environment that crops planted outside its rainforest home simply do not produce fruit.

Trees add a vast wealth of nutritious and delicious ingredients and flavours to our diet – our lives would be so much poorer and duller without them.

FEASTING AND CELEBRATING

Brazil nut

Bertholletia excelsa

Brazil is the most biodiverse country on Earth. It is home to 46,000 species of plants and fungi, of which 19,500 grow nowhere else, with a vast diversity of orchids, palms, hardwood trees and crops, including the Brazil nut tree, *Bertholletia excelsa*. This towering tree is one of the tallest of the Amazonian rainforest, growing up to 50–60 metres (165–196 feet) high, and is found not only in Brazil, as its name would suggest, but also in Bolivia, Peru, Venezuela, Amazonian Colombia and the Guianas. Although this species has fine timber, it is the seeds of the Brazil nut tree that are most prized, and are harvested and exported. Full of protein, carbohydrate and fats, and a good source of minerals, they are in great demand worldwide. Brazil nuts are one of the most valuable non-timber crops of the Amazon, and are still gathered from the wild, forming a valuable source of income for local people. Most of the Brazil nuts we eat in fact come from Bolivia, rather than Brazil.

The familiar angular seeds or nuts develop inside a heavy, woody, round fruit capsule ... which falls with great velocity once ripe.

The familiar angular seeds or nuts develop inside a heavy, woody, round fruit capsule, weighing up to 2 kilograms (almost 5 pounds), which, after over a year developing on the tree, falls with great velocity once ripe. Each fruit capsule holds between 12 and 25 seeds, tightly packed inside. The main disperser of these seeds is the large forest rodent species known as the agouti, which has sufficient determination and teeth sharp enough to gnaw through the tough outer casing of the fruit to reach the nutritious seeds within. Agoutis have the habit of eating some seeds and caching the rest by burying them; they then sometimes forget where they have left them, thereby aiding the perpetuation of the trees. If buried in a spot where there is enough light, the seeds will eventually germinate and grow up into gaps in the forest.

Brazil nut trees only produce good crops of fruit in their native rainforest habitat, where the intricate ecological partnerships they need for pollination exist.

OPPOSITE AND BELOW The individual angular Brazil nuts are packed tightly inside large, heavy spherical fruits which drop to the forest floor. Only the agouti has teeth sharp enough to gnaw holes in the fruit to get to the nuts inside.

One of the most extraordinary things about the Brazil nut is its refusal to be domesticated. These trees cannot be grown with any great success in plantations and the nuts have to be collected from the wild. This has everything to do with the tree's wonderful and convoluted pollination story. Each of Brazil's biomes has a complex ecology where many species have evolved to depend on one another. Species become vulnerable when these relationships are threatened. Sir Ghillean Prance, a former director of the Royal Botanic Gardens, Kew, and an authority on rainforest ecology, realized after years of research that the Brazil nut harvest in the Amazon was reliant on the health of a web of surrounding plants and animals. The tree itself depends almost entirely on large-bodied euglossine or orchid bees to pollinate its robust creamy-white flowers in the early hours of the morning. The female bees will only mate with males that have successfully gathered a cocktail of scented wax from various orchid species that grow nearby, including *Coryanthes vasquezii*. If the orchids are removed during logging or through other human activities, the bees disappear too; the Brazil nut flowers then remain unpollinated and the Brazil nut harvest fails.

Because the tree is so dependent on this natural web – from bee to agouti – it is difficult to farm, but the nuts are also difficult to harvest sustainably from the wild too. A number of scientific papers have shown that *Bertholletia excelsa* is not regenerating properly in its natural range, as too many fruits are being taken and therefore older trees are not being replaced naturally by younger ones in sufficient numbers to ensure future reliable harvests. The International Union for the Conservation of Nature has classed the Brazil nut tree as a species vulnerable to extinction. Its natural habitat, the Amazon Basin, is a reservoir of biodiversity and home to many other economically important trees such as cacao and rubber, as well as species of great economic potential and medicinal value.

FEASTING AND CELEBRATING

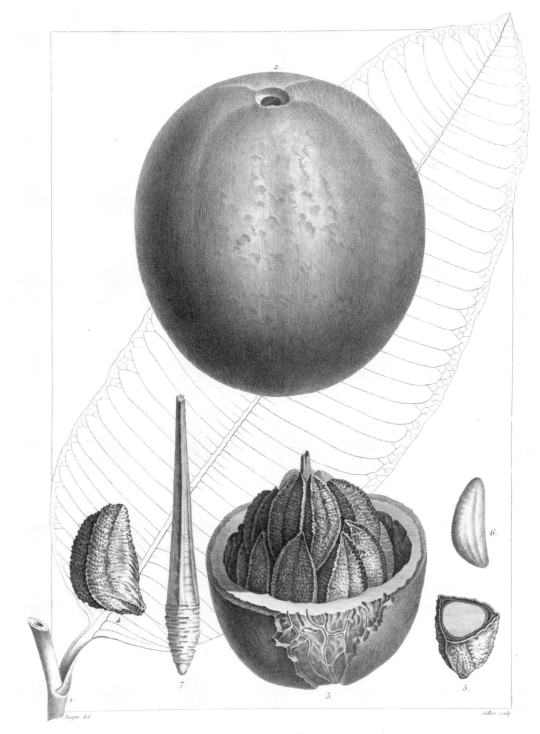

2.

4.

1.

7.

3.

6.

5.

Turpin del.

Selber sculp.

BERTHOLLETIA excelsa.

De l'Imprimerie de Langlois.

Cacao

Theobroma cacao

Every chocolate-lover owes a deep debt of gratitude to the unassuming tropical tree species, *Theobroma cacao*, but also to the ancient peoples who first discovered its delights. The cacao tree from which chocolate is made was expertly cultivated by the ancient Maya and Aztec peoples of Central America, but recent research suggests that its first use dates back to much earlier societies living in South America, in Ecuador, over 5,000 years ago. Studies of ceramics of the Mayo-Chinchipe culture show cacao was being consumed there, most probably as a drink. It's now thought that its popularity spread, through trade, up into Central America over hundreds of years.

The Mayan name for the plant was *kakaw* or *kakawa* – which has become cacao – now both its scientific species name and its common name. It was eighteenth-century taxonomist and chocolate-lover Carl Linnaeus who gave the genus its name *Theobroma* – meaning 'food of the gods'. Cacao is a member of the Malvaceae or mallow family and is related to hibiscus and okra. Believed to have originated in the rainforests of the Amazon, cacao is a small tree that grows up to 8 metres (26 feet) tall. It needs a tropical climate, but also shade and high humidity, preferring to grow in the understorey of the rainforest. As a crop it therefore has to be cultivated under taller trees. Today, these include other crops such as bananas (*Musa*) or rubber (*Hevea brasiliensis*), giving added benefit to the farmer.

Cacao produces surprisingly small and delicate-looking pink flowers directly from its trunks (a feature known as cauliflory) throughout the year. Once these are pollinated by a species of midge, the large pods develop. These yellow or red, ridged oval fruits (technically berries) grow up to 25 centimetres (10 inches) long, and are filled with a sweet white pulp that holds thirty to forty seeds or 'beans'. In a good year, a healthy tree can produce around thirty such pods.

This eighteenth-century depiction of cacao pods shows how each fruit is packed with sweet pulp enclosing the much sought-after beans.

Cacaos, Cacavifera,
Chocolat • Mandel.

Pl.62.   Page 101.

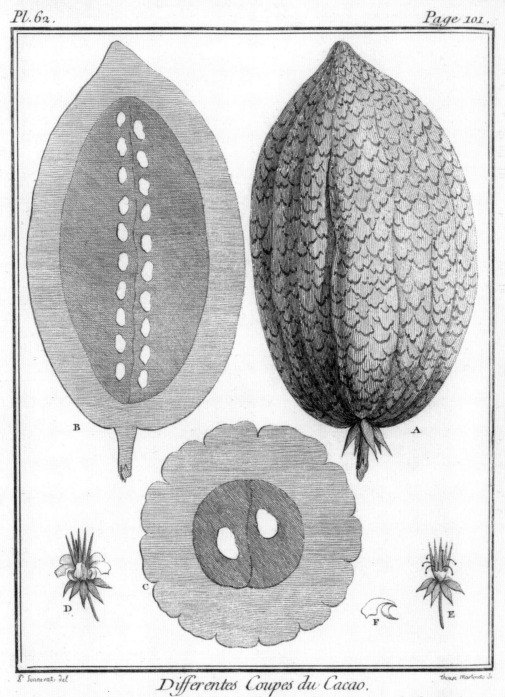

L. Sonnerat, del.   Thenze Martinet Sc.

Differentes Coupes du Cacao.

A . Le Cacao . B . Coupe perpendiculaire du Fruit . C . Coupe horizontale du Fruit
D . la Fleur vüe à la Loupe . E . Developpement de la Fleur vüe à la Loupe
E . une des Petales vüe à la Loupe .

The Maya were well practised in how to grow cacao in the forest, and how to ferment, toast, dry and grind the beans to make a paste. This was then mixed with hot water and poured from a height from one vessel into another to produce a foaming drink. Later, the Aztec people also treasured the cacao tree, believing it to be a gift from the gods, as its current scientific name reflects. The Aztecs are thought to have preferred their version of the drink cold, and prepared it using high value implements, reputedly drinking it from golden cups. This was a truly special drink – only for the elite of their society and for warriors. Commoners, women and children were not allowed even to taste it. The beans were precious and highly prized: they were traded widely and even used as a form of currency. Cacao was also offered in both tribute and as a sacrifice. Moctezuma (Montezuma) II – the ruler of Tenochtitlan (modern Mexico City) between 1502 and 1520, was recorded as drinking many flavoured versions of the foaming dark drink they called *cachuatl*. Vanilla, chilli, spices, honey and herbs or flowers were all used to spice up his chocolate drink, to which he seems to have been slightly addicted.

Always thought to be delicious and desirable, dark chocolate and pure cacao are now known also to have real health benefits. Cacao contains phenols and flavonoids which have antioxidant effects thought to inhibit cancer and cardiovascular disease.

Brought back to Europe by the Spaniards around 1544, cacao soon became a novel drink at the Spanish court. It was then introduced throughout Europe in the sixteenth century, and was originally taken as a medicinal drink to aid digestion and to settle the stomach. It was when it was once more mixed with hot water, as the Maya had done, that the name 'chocolate' was born and this drink grew in popularity. It was much favoured by French royalty at Versailles – Louis XV reputedly had his own recipe. In the seventeenth century chocolate houses, like today's coffee shops, sprang up in Oxford and London. Samuel Pepys recorded in the 1660s that he often enjoyed a morning drink of 'chocolatte'.

Another important development in our love affair with chocolate came after Sir Hans Sloane visited Jamaica and was distinctly unimpressed by the Central American way of drinking chocolate, considering it 'fit only for swine'. He then devised a recipe using hot milk and sugar that made it immensely more palatable and desirable. Chocolate grinding and making began to become an industry across Europe in the eighteenth century and in 1828 the Dutchman Coenraad van Houten invented a process that created cocoa powder and also facilitated the creation of solid chocolate. By 1842 Cadbury brothers were in business in the UK selling powdered and solid chocolate made from

Cacao pods, and the beans they contain, were once so important and desirable they were even used as currency.

OPPOSITE A fruiting cacao
tree, ready for harvesting.
The small flowers and large
pods that follow pollination
are produced direct from the
trunk and branches.

BELOW The Economic
Botany Collection at the
Royal Botanic Gardens,
Kew, holds a wealth of plant
products, including these
rolls of pure cacao from
Trinidad and Tobago.

cocoa butter and ground beans, but it remained a luxury item until
the mid-nineteenth century when import levies were lifted. Swiss
chocolatiers also began to create new and wonderful confections, and
demand and production took off.

Always thought to be delicious and desirable, dark chocolate and
pure cacao are now known also to have real health benefits. Cacao contains phenols and flavonoids which have antioxidant effects thought
to inhibit cancer and cardiovascular disease. Cacao also contains theo-
bromine and caffeine alkaloids. It is these that are believed to improve
mental alertness and can have an addictive effect.

Today, chocolate is enjoyed across the world in myriad forms and
is generally sold at an affordable price. Over 4 million tonnes of beans
are now produced each year and it is predicted that demand will soon
outstrip supply. Although native to tropical America, today most cacao
is grown in West Africa – with the Ivory Coast and Ghana being the
top producers. It is an extremely important crop for around 5–6 million small farmers across the tropics.

But there is a potential threat to the security of our much-loved
chocolate crop. Unfortunately, *Theobroma cacao* has limited genetic variability and seems to have little natural resistance to pests and diseases.
Plantations are plagued by problems including fungi such as frosty pod
rot and witches' broom in the Americas, swollen shoot virus in Africa,
and cocoa pod borer in Southeast Asia. Along with climate change and
the poverty of the regions in which cacao is often grown, and with so
many people depending on it for their livelihoods, the stakes are high.
However, many scientists are working together to find solutions to save
the cacao tree. In 2010 the DNA of the highly prized ancient Maya
variety 'Criollo' was fully sequenced, allowing researchers to discover
the genes responsible for protecting the plant against disease and so
help breed hardier trees. Conserving the wild relatives of *T. cacao* in
their natural habitat may also offer valuable genes
to safeguard the future of our favourite sweet treat,
another very good reason to support the protection
of the rainforests.

CACAO

Cinnamon

Cinnamomum verum

════════

The warm, delicate aroma of cinnamon is indicative of the climate in which it naturally grows, for cinnamon as a spice comes from a tropical tree, *Cinnamomum verum*. Although it is now cultivated in many tropical parts of the world, including India and Bangladesh, and also Brazil and Jamaica, the tree is native to Sri Lanka. Most of our true cinnamon still comes from the island and is regarded by connoisseurs as having the best flavour.

Cleopatra included cinnamon, along with gold, silver, emeralds and pearls, as the most valuable of the royal treasures she gathered together for her tomb.

When left to grow naturally the cinnamon tree can reach a height of 7–10 metres (23–33 feet), but in cultivation it is pruned and coppiced to keep it to around 3 metres (10 feet) tall to make it easier to harvest. An evergreen species, it has deeply veined glossy leaves (red when young), which have a spicy smell if crushed, but it is the inner bark of the young, green-orange branches that is used to make the spice. Once the outer bark is scraped away, the thin inner bark is peeled off and cleaned, with the resulting strips left to dry. As they do so, they roll up into tubes, which are then cut into the familiar quills. The spice is also sold in powdered form, and an oil is distilled from the leaves and bark.

There are over two hundred species of *Cinnamomum*, but *verum* is considered to yield the most delicate spice. It has been exported from Sri Lanka for hundreds of years, as reflected in its previous scientific name, *C. zeylanicum*, 'from Ceylon', while its current name means 'true cinnamon'. A close relation is *C. cassia* – known as Chinese cinnamon or cassia. The spice made from this species has an even longer history, and in antiquity cassia from China was traded along the Silk Roads. It has a stronger taste than true cinnamon and is now generally cheaper due to its abundance. Still today powdered cinnamon is often

The thin inner bark of young branches of *Cinnamomum verum* is peeled away and then dried. As it dries it naturally curls into the quills we are familiar with.

mixed with cassia, and many baked 'cinnamon' goods are actually made with cassia.

Over the thousands of years they have been traded, cinnamon and cassia have been employed in many ways. Classical authors such as Herodotus and Pliny often mention 'cinnamon', though it is hard to be certain which of the two spices is actually being referred to. It also appears several times in the Old Testament of the Bible with other fragrant spices, and was a favourite of the ancient Greeks and Romans. Plutarch records that the Egyptian pharaoh Cleopatra included cinnamon, along with gold, silver, emeralds and pearls, as the most valuable of the royal treasures she gathered together for her tomb. In antiquity 'cinnamon' was a costly item and not a flavouring for food – it was used for incense and as an aphrodisiac and also as a medicine for all manner of ailments, as it still is in places today. The ancient Egyptians used it as one of the ingredients for mummifying the dead.

The early trade in the spice was monopolized by the Arabs, but by the Middle Ages true cinnamon was being imported into Europe via the merchants of Constantinople and Venice as part of the highly lucrative Spice Trade. European nations competed to discover the origin of this and other costly spices and so break the trade monopoly. In the late fifteenth century, Portuguese adventurers exploring maritime routes located the source of cinnamon in Sri Lanka. The Dutch East India Company later created their own monopoly on trade in the spice.

Today, we still enjoy the flavour of this tropical tree's bark in many dishes and drinks, both sweet and savoury, from apple pies and pastries to curries, tagines and mole sauce. The particular taste of cinnamon comes from a volatile oil in the bark, which contains numerous chemical compounds. The most aromatic of these is cinnamaldehyde, which also shows bactericidal powers and it is suggested it may possibly have acted as a food preservative in times before refrigeration. Research has shown that cinnamon oil does have some anti-microbial properties and could also be beneficial for lowering blood sugar, cholesterol and blood pressure, but more study is required. *Cinnamomum* continues to be used in traditional Chinese medicine for a whole range of conditions, from relieving nausea and digestive problems, to easing colds and fever, diarrhoea and gynaecological problems.

L'image de l'arbre qui produit la Canelle.

Palmae
(Cocoineae)

Taf. II. Cocos nucifera L.

Coconut

Cocos nucifera

=====

The very word 'coconut' conjures up images of tropical islands, with waving palms fringing white sandy beaches at the edge of a turquoise sea. In popular imagination, coconuts are seen as the typical, ubiquitous palm of the tropics, and they have been planted in many places across such regions because of their wide-ranging usefulness – for food, drink, medicines, building materials, clothing and many other products. Appropriately, the coconut is known as the 'tree of life' in the Philippines and it is grown commercially in at least eighty countries.

It is no longer found in the wild and is said to be the most naturally widespread fruit plant on the planet.

Most botanists think that the coconut is native to somewhere in the southwestern Pacific, but as it is no longer found in the wild and is said to be the most naturally widespread fruit plant on the planet it is impossible to pin down its exact origins. In the late thirteenth century the famed traveller Marco Polo encountered coconuts and summed them up very accurately: 'The Indian nuts also grow here, of the size of a man's head, containing an edible substance that is sweet and pleasant to the taste, and white as milk. The cavity of this pulp is filled with a liquor clear as water, cool, and better flavoured and more delicate than wine or any kind of drink whatever.' Many sailors and travellers of the eighteenth century grew to admire and rely on the coconut in their explorations.

In scientific terms neither a true tree nor a nut, the coconut is a palm (of the Arecaceae family), which in the botanical sense are herbaceous plants. Their tree-like form consists of a tall fibrous flexible stem (growing up to around 25 metres or over 80 feet high) topped by leaves 4 metres (13 feet) long which are divided into the familiar 'fronds'. The pliant stems and shallow roots help the trees to survive strong winds along coastlines. They are extremely versatile plants and

The coconut palm's brown 'nut' is the stone of a much larger fruit. The white flesh and sweet water inside are food for a developing future new plant.

grow in places such as salty, sandy shores which other palms and trees cannot tolerate. As for the brown 'nut' we are familiar with, this is actually the stone of the bigger fruit (or drupe) of the palm. The green fibrous outer husk is removed (and used as coir) in order to extract the coconut inside, but in nature it serves a vital function for the seed as it provides buoyancy as the fruit is carried out to sea by the tides, helping it to travel great distances along coastlines. A coconut can survive in salt water for around two months, in which time it could theoretically travel an incredible 5,000 kilometres (3,100 miles) on ocean currents. Only once it is washed up and near fresh water will the seed germinate.

Within the coconut is the embryonic plant: the white 'flesh' is actually the kernel of the seed, which hardens as the seed ripens, while the liquid that develops inside becomes sweeter. The three characteristic 'eyes' at one end of the nut are pores through which a germinating shoot can emerge. The kernel and liquid are both food for the developing shoot.

The refreshing coconut water offers a safe and nutritious drink in the tropics and is said to be rich in electrolytes, though drinking it in excess can act as a diuretic and it is also possible to take in too much potassium from this harmless-looking liquid. Coconut water is not to be confused with coconut milk, an ingredient used in countless recipes,

which is made from grated coconut flesh. Coconut oil, meanwhile, is made from pressing dried coconut flesh. This is also becoming increasingly popular for cooking in the West, and is an ingredient for a whole range of products including margarines, baked goods and sweets, as well as cosmetics such as moisturisers and soaps.

What may be more of a surprise is the fact that the coconut also has medicinal properties. Scientific studies have looked into its active molecules, and these have been shown to possess many beneficial effects – from protecting kidney, heart and liver functions to having analgesic, anti-inflammatory and bactericidal effects. Several parts of the plant – fibres, leaves and milk – have a role in traditional medicines for diarrhoea, and the oil has also been used to treat skin conditions, burns and other wounds. With new research it is possible that the coconut could become a source of low-cost medicines around the world, adding to the many and varied uses of this extraordinary plant.

Nutmeg

Myristica fragrans

It took years of intrigue, bravery, deception and bloody war in the sixteenth and seventeenth centuries for Europeans to gain direct access to this fragrant spice. Thousands lost their lives in the fight to control its trade, and it was once worth more than its weight in gold. The aromatic spice we call nutmeg comes from the fruit, or drupe, of the evergreen tropical tree *Myristica fragrans* (meaning myrrh-like fragrance). A relatively large tree, reaching up to around 20 metres (66 feet) in height, *Myristica* is slow growing but flowers continuously and can yield up to 20,000 fruits a year. It is native to the Banda Islands in the province of Maluku, Indonesia, where it thrives in the deep moist volcanic soil of these few, small 'Spice Islands'. Until the mid-1800s these were the only places where nutmegs were commercially grown and traded.

To Europeans in the fifteenth century the source of nutmeg was a complete mystery.... Wild myths grew up around its exact origin.

The fruits of the nutmeg tree develop from its tiny yellow bell-shaped flowers. Male and female flowers grow on separate trees (meaning the tree is dioecious), so both are needed for pollination to occur. The fruits that then form are rounded and pale gold in colour, resembling apricots. The outer aromatic fleshy skin is edible and can be made into jam or candied as a sweet or dessert, but it is the kernel inside that yields the spice. Ripe fruits split open to reveal the precious oval nutmeg inside. Surrounding this hard brown 'nut' is a bright crimson lacy aril, or seed covering, from which we obtain the second spice of the nutmeg tree – mace. This is separated and dried to make that similar but more delicate flavouring. An essential oil can also be distilled from ground nutmeg, which is used today in foods, drinks, perfumery, cosmetics and in some pharmaceuticals such as cough medicines.

In the early seventeenth century, when this illustration was drawn, the race to find and control the source of nutmeg was at its height.

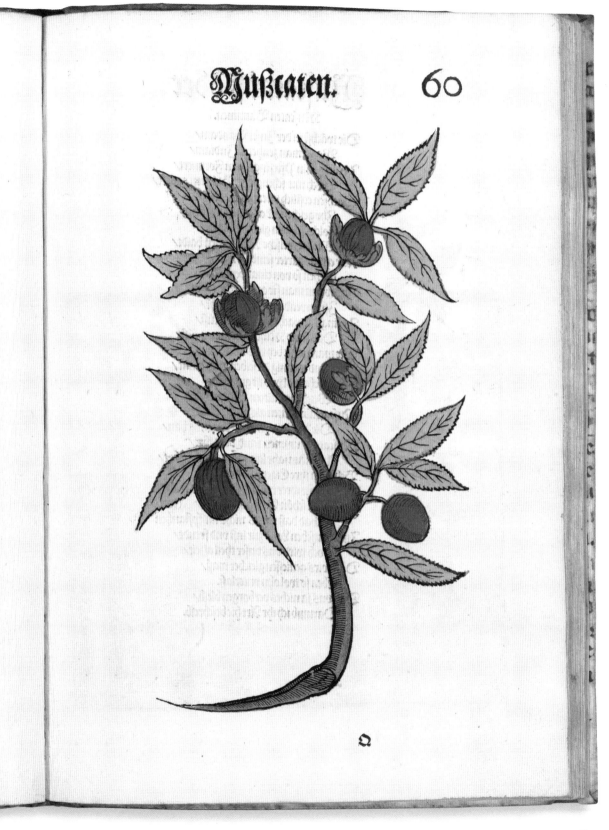

OPPOSITE Ripe fruits split open to reveal the oval precious nutmeg inside. Around the hard brown kernel is a lacy crimson aril, which is separated and dried to produce mace, with its similar but more delicate flavour.

BELOW Nutmeg has been prized and traded for many hundreds of years. It was known to the Romans, used in India and China, and was described by Marco Polo in the thirteenth century, though the home of the nutmeg tree was a mystery for centuries.

To Europeans in the fifteenth century the source of nutmeg was a complete mystery. It had been traded for centuries, and at the end of the thirteenth century the Italian explorer Marco Polo had described nutmegs as among the 'surpassing wealth' of Java. Spices were bought from merchants in Venice, who in turn had acquired them from Constantinople and a chain of other merchants stretching from the mysterious East. Wild myths grew up around its exact origin. By the sixteenth century nutmeg had become highly prized both as a food flavouring and a valued medicine and preservative. Because of its ever increasing price, many sought to find the land where the nutmeg grew. As with cinnamon (p. 78), the Portuguese won the race, reaching the Banda Islands and taking over the port of Malacca in 1511. They filled two ships with nutmeg, mace and cloves, which they later sold for approximately one thousand times what they had paid the local people.

Nutmeg was used to treat all manner of ailments, from diarrhoea to digestive problems and more besides. When Elizabethan doctors recommended nutmeg in pomanders to ward off bubonic plague, its value rose even further and it became one of the most sought-after trading items in the world. The lure of such immense wealth drew the British into the spice trade, determined to gain their own share of the profits. In 1603 an expedition led by James Lancaster landed on the tiny island of Run (or Rhun), a small atoll 16 kilometres (10 miles) from the main Banda Islands. Here there were plentiful nutmeg trees, and after the British had made a successful first impression with the local people, trade commenced. However, the Dutch held the monopoly on spices in the region and would not tolerate rivals in this highly lucrative business. They occupied the Banda Islands by force in 1621, massacring its people and expelling survivors or selling them as slaves. They created nutmeg plantations, and in order to maintain their monopoly the Dutch East India Company destroyed any nutmeg trees not under their control. The British later ceded the island of Run to the Dutch in 1667 in return for the little-known island of Manhattan in North America. The Dutch monopoly did not survive long, however, as British explorers took nutmeg trees to Sri Lanka and other colonial tropical areas under their control. In the eighteenth century the rather aptly named French botanist Pierre Poivre made several attempts at great risk to smuggle nutmeg seeds or plants out of the Banda Islands, and

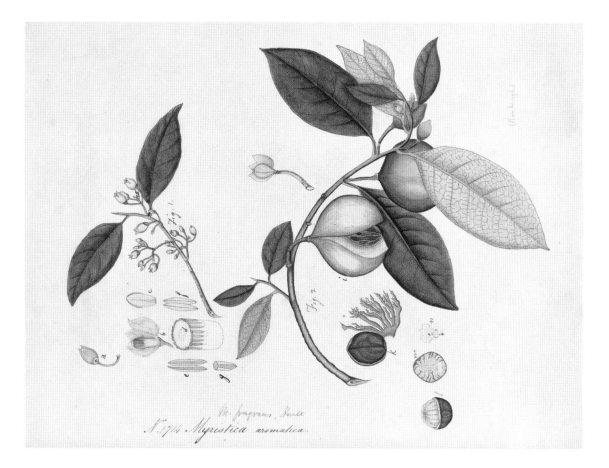

finally succeeded in establishing some trees in Réunion and Mauritius in the Indian Ocean. Nutmeg-growing spread around the globe, and today the island of Grenada in the Caribbean is one of the largest producers of nutmegs in the world.

The culinary qualities and virtues of this hard-won eastern delight continue to be valued, and myriad recipes exist across several cuisines to make use of its aromatic, warming flavour. Nutmeg is also a source of minerals, B vitamins and antioxidants, and is being investigated for its potential medicinal properties, including possible protection against certain potent bacteria, as an aid to liver function and also an antidepressant. Although some people take nutmeg medicinally, excessive over-indulgence can have severe consequences including allergic reactions, hallucinations and even death, and it is also highly toxic to dogs.

The stirring tale of the nutmeg is another example of how the trade and movement of seeds around the world has affected the lives of millions and the economies of entire nations.

Olive

Olea europaea

───

A tree that featured in a contest between Olympian deities, the European olive has played an important part in human history. According to Greek myth, in the dispute between Athena and Poseidon to decide who would be the patron deity of Greece's principal city, the goddess planted an olive tree sapling. Athena's gift was so useful to the inhabitants that she was the victor, and hence the city became known as Athens. So highly esteemed was the olive that its oil was presented as a prize to the winning athletes in the ancient Athenian games, and an olive tree still grows on the Acropolis as a symbol of this foundation story.

A wreath including olive leaves was found in the tomb of the Egyptian boy-king Tutankhamun, dating back 3,300 years.

The sheer usefulness of the hard wood, fruits and oil of the olive tree have made it an extremely valuable commodity over thousands of years. In addition, the olive also has great symbolic significance, epitomizing all the blessings of life – longevity and fertility, nourishment, hope, wisdom and wealth. The olive branch as an emblem of peace has ancient roots, perhaps best known from the story of Noah and the Ark in the Bible, and is still universal today. Olive branches are represented on the flag of the United Nations, where they encircle a map of the world, and also on the Great Seal of the United States, as well as on the flags of Cyprus and Eritrea. A small golden olive branch was left on the Moon by astronauts of NASA's Apollo 11 mission, representing a wish for peace for all the peoples of Earth.

Olea europaea, its subspecies and numerous domestic cultivars all thrive in the baking sun and dry heat of the Mediterranean, where they often grow on thin limestone soils. These small but tough evergreen trees, which are rarely over 10–15 metres (33–50 feet) tall, have narrow waxy leaves that help retain precious moisture. Olive trees can

The olive tree is synonymous with landscapes around the Mediterranean, where many ancient groves still thrive even on the poorest soils and with little rainfall.

withstand severe stress and even wildfires, often re-growing from the base if not too badly damaged. Due to the harsh environment, olives grow slowly but can be fruitful for a relatively long period of time. A tree in Brijuni National Park in Croatia radiocarbon-dated in the 1960s as 1,600 years old was still yielding around 30 kilograms (66 pounds) of olives each year. Many groves are reputed to be centuries old, while some trees, for example in Italy, Greece, Malta and Croatia, are claimed to be as much as 2,000 years old. One gnarled but fruitful tree in Crete is even estimated to be over 3,500 years old, but because it is hollow there is no heartwood that can be dated by tree rings. In the Garden of Gethsemane in Jerusalem truly ancient-looking olive trees have been radiocarbon-dated and found to be mainly from the twelfth century and so may have been planted at the time of the Crusades. Precisely when ancient people discovered the merits and gifts of the olive tree, however, is lost in the mists of time. Archaeological excavations in Israel have reportedly uncovered olive stones that suggest around 20,000 years of use.

Olive trees are widespread across their native Mediterranean, and are also found in North Africa and across into Asia, and have been taken to other parts of the world far from their original habitat. The

olive was introduced into California in the United States in the late eighteenth century by Spanish missionaries, and as the climate there is similar to the Mediterranean, trees thrived and cultivation took off in the nineteenth century. Olive groves now cover around 14,000 hectares (35,000 acres) and one cultivar that was developed there is known as the 'Mission olive'. Olives are now also grown in South Africa and Australia, where production of oil is increasing, and even in Rajasthan in India.

With such a long history of cultivation, there are now over a thousand cultivars of the olive, and six distinct subspecies of *Olea europaea* occur in separate regions of its native range. The famous fruits (technically known as drupes) develop from tiny white flowers, borne in clusters, which are wind pollinated. As the fruits mature in the late months of autumn they turn from green to black. Olives cannot be eaten straight from the tree, however, because of their bitterness, caused by phenolic compounds such as oleuropin. After being harvested by hand to avoid bruising they must be soaked in water and then cured in brine for several days before being rinsed again.

The best oil is cold-pressed from the raw fruits. This 'extra virgin oil' has the lowest acidity and highest purity. In addition to being nutritious and having possible health benefits, the oil has been used for all manner of purposes down the ages – from anointing and blessing kings, priests, athletes and sacrifices, to fuelling lamps, making medications, preserving food and for massaging and cleansing.

But the uses of this valuable tree don't end there, for the hard and durable wood is also highly sought after for carving and furniture

ABOVE The olive branch as a symbol of peace has a long history. It famously features in the story of Noah in the Bible, when a dove that was released after the great Flood returned with a fresh olive branch, representing hope and the promise of peace.

OPPOSITE Old olive groves provide valuable habitats for wildlife and support an abundance of wild flowers. They have also inspired poets, philosophers and artists.

Olive trees are slow growing but continue fruiting for a relatively long period of time. The oil has been valued for a multitude of purposes over thousands of years, and the Greek author Homer referred to the first oil pressed from olives as 'liquid gold'.

– Homer describes the bed of Odysseus as being made from a still rooted olive – while the trees can be planted for welcome shade, as fire-breaks and to control soil erosion. Olive groves create valuable habitats in arid regions for wildlife, and are often filled with a riot of wild flowers in spring. Philosophers, poets and painters have been inspired by orchards of olives. Vincent van Gogh completed at least fifteen paintings of ancient olive trees while recovering at the asylum in Saint-Rémy-de-Provence in southern France. He was keen to capture the spirit of the natural Provençal landscape and gained peace and solace from it. In a letter to his brother Theo he wrote that he found the 'rustle of the olive grove has something very secret in it, and immensely old'.

The collections of the Royal Botanic Gardens, Kew, contain an astonishing range of products made from various parts of the olive tree, including walking sticks, smoking pipes, spoons and rosaries, but none are more poignant and evocative perhaps than a wreath including olive leaves found in the tomb of the Egyptian boy-king Tutankhamun, dating back around 3,300 years.

The future of this important crop is a matter of some concern as climate change is predicted to result in higher temperatures and increased droughts in the Mediterranean Basin, and the fatal bacterium *Xylella fastidiosa* is on the march across the olive groves of southern Europe. Some areas of olive production may become completely unviable. But our special and close relationship with the olive, culturally and economically, will surely endure.

FEASTING AND CELEBRATING

Pecan

Carya illinoinensis

In its natural home along the banks of large rivers of the southern United States of America and into Mexico, where it flourishes in deep, rich soils, the pecan develops a long tap root from an early age and is fast growing. It then makes a large, graceful deciduous tree reaching up to 40 metres (130 feet) tall, often with a trunk of up to 2 metres (6½ feet) in diameter. It has feathery compound leaves made up of numerous leaflets and male catkins that appear in spring. After the small female flowers are pollinated, the desirable nuts form, encased in a husk which characteristically splits into four when ripe.

Because of the wrinkled look of the kernel when removed from its thin shell, they called it the nuez do la arruga, *which translates as 'wrinkle nut'.*

The pecan is one of around eighteen species of *Carya* or hickory native to North America, with at least one species, *C. cathayensis*, found in Southeast Asia. The name for the genus comes from the ancient Greek *karyon* for 'nut kernel', and the species *Carya illinoinensis*, the true pecan, was most probably first described from a specimen collected in the state of Illinois in the midwestern region of the United States, where it was known as the Illinois nut.

For many indigenous American peoples in different parts of the tree's large natural range the nuts were a precious food source and an important trade item. The word 'pecan' is from Algonquian and means 'nuts that need a stone to crack', so it encompasses all the walnuts and the hickory nuts including the pecan. Pecans have a rich buttery taste and approximately 33 pecan halves (66 whole nuts) will provide 690 calories and over 100 per cent of our daily total for fat. They are also a rich source of dietary fibre, manganese, magnesium, phosphorus, zinc and thiamine, so in a nutshell they are generally good for our diet, eaten raw or cooked.

The pecan was encountered by Europeans when the Spanish explorers discovered the trees growing in Louisiana, Texas and Mexico. Because of the wrinkled look of the kernel when removed from its thin shell, they called it the *nuez do la arruga*, which translates as 'wrinkle nut'. Early settlers began planting the tree and took it to other parts of the United States; Thomas Jefferson grew pecans at Monticello in Virginia. A trade in the nuts developed and they began to be exported to other parts of the world. However, there is a great natural variability in growth habit and flavour among wild pecans, which is a disadvantage for successful commercial cultivation. In addition, the tree's long tap root means it is very difficult to transplant, and for good nut production several trees are needed for cross-pollination.

Attempts at grafting to obtain the desired results met with little success until an enslaved gardener named Antoine grafted a wild pecan with the right characteristics on rooting stock and produced a cultivar of pecan known as 'Centennial'. Then in 1874 an English cabinet-maker, Edmond E. Risien, who had a fascination for the pecan, moved to San Saba County in central Texas to work on the cultivation and breeding of the native pecans. From a plate of prize-winning pecans at a local pecan show he tracked down a prolific fruiting tree that had had its main branches badly cut in order to collect the nuts. To save the tree from being cut down, Risien bought it with the surrounding land and founded the West Texas Pecan Nursery. The tree soon recovered and developed a new crown and began producing nuts again. He used the fruits from this large specimen tree to plant a 16-hectare (40-acre) commercial orchard, but the trees failed to produce nuts as good as the ones from the mother tree.

To improve yields and quality Riesen travelled around the state of Texas looking for good potential father trees, collecting male flowers and using their pollen to artificially cross-pollinate with the mother tree. As a result, he produced many new prolific-fruiting varieties that are still grown today in commercial pecan orchards, such as 'San Saba Improved', 'Texas Prolific', 'Liberty Bond' and 'Western Schley', single trees of which can produce around 450 kilograms (1,000 pounds) in a good season. The tree that Risien saved from the axe is now called the 'San Saba Mother Pecan', a notable tree in Texas. San Saba has been

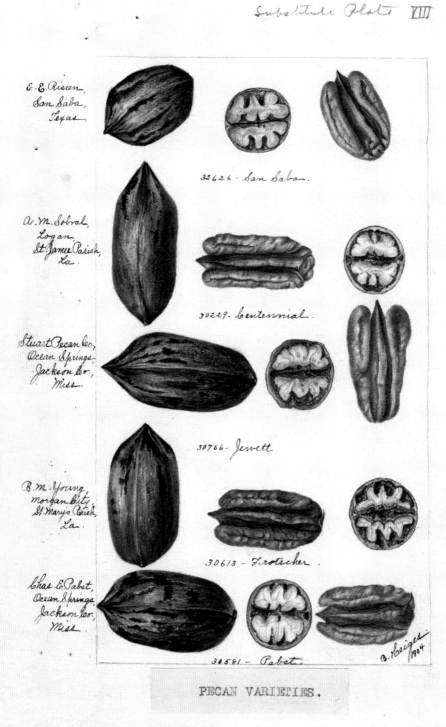

E.E.Risien,
San Saba,
Texas.

32626 - San Saba.

A. M. Sobral,
Logan
St. James Parish,
La.

30229. Centennial.

Stuart Pecan Co.,
Ocean Springs,
Jackson Co.,
Miss.

30766 - Jewett

B. M. Young,
Morgan City
St. Mary's Parish,
La.

30613 - Frotscher.

Chas E. Pabst,
Ocean Springs
Jackson Co.,
Miss.

30581 - Pabst.

B. Heiges.
1904

PECAN VARIETIES.

Pacanenut Hickory.
Juglans oliver-formis.

The pecan tree is one of around eighteen species of hickory. It has feathery compound leaves made up of numerous leaflets. Following pollination the flowers produce the nuts, each encased in a husk which splits into four when ripe.

proclaimed the pecan capital of the world, and in 1919 the pecan tree was officially made the state tree of Texas.

It is not only the nut of this beneficial family of trees that is an important commercial product – the timber of the hickory is also used for furniture and flooring, and is the wood of choice in the US for smoking meats and fish to add flavour. As it is very hard, rigid, dense and shock resistant, it also makes an ideal wood for the handles of axes, mattocks and hammers, and the shafts of sports equipment including lacrosse and hurley sticks and golf clubs; the earliest baseball bats were made from hickory, before preference switched to ash or maple.

Japanese persimmon

Diospyros kaki

The name *Diospyros* is derived from the Greek words meaning divine fruit or wheat, giving some idea of the regard this tree is held in. *Diospyros kaki* sits in a large genus of between 500 and 700 species of deciduous and evergreen trees and shrubs, which in turn belongs to the Ebony family (Ebenaceae), which also includes ebony itself, *Diospyros ebenum* (p. 152).

Diospyros kaki is generally known as the Japanese persimmon or *kaki*, hence the same specific epithet. Other common names include Chinese or Oriental persimmon. Originally native to China, where it has a history of cultivation stretching back over 2,000 years, and also Korea, it has also been grown and cherished in Japan for centuries. In the nineteenth century, it was introduced further afield, to California, Brazil and southern Europe.

> *The dried seeds were made into buttons for the uniforms of soldiers in the American Civil War.*

In late autumn, the oval leaves of this small tree colour beautifully and then fall, leaving the large tomato-like fruits, up to 10 centimetres (4 inches) in diameter, called 'simmons', hanging from the branches to ripen to a rich shiny orange. In Japan the fruits are strung up after harvest and suspended on lines around the lower eaves of traditional Minka houses. They are allowed to dry naturally in the sun and then eaten as high-energy sweets called *hoshigaki*. In China they are peeled, air dried and squashed flat to make edible snacks or for use in cooking.

There are two main types of these edible fruits, astringent and non-astringent, which are also slightly different in shape. The astringent cultivars, known in Japan as *hachiya*, are more oval and pointed, and are high in proanthocyanidin tannins, which means they must be allowed to ripen fully on the tree until soft before eating or they will dry out the mouth with a single bite, never to be tried again.

Native to China, where they have been cultivated for thousands of years, persimmons are known as *kaki* in Japan. The fruits, rich in vitamins and fibre, are divided into two main types: non-astringent and astringent.

BELOW The large, tomato-like persimmons ripen in autumn to a rich, shiny orange. The leaves also colour beautifully and fall, leaving the glowing fruits hanging on the bare branches.

When fully ripe, the flesh matures into a thick pulpy puree contained in a thin, shiny yellow or orange waxy skin. These are generally the kind that are used for drying. The non-astringent type, *fuyu gaki*, can be eaten unripe, when they are firm and crisp, or left to ripen to a jelly with a rich sweetness. So why eat an astringent persimmon? Well, when ripened it is said to have a far superior flavour to the non-astringent cultivars. In the Far East *kaki* can be served as a dessert after a meal, and dissecting the fruit into sections is an art form in itself.

Fresh or dried, persimmons have a soft to fibrous texture, with twice as much dietary fibre as an apple, and are high in vitamins A and C, potassium, manganese, copper and phosphorus. Like most other commercial fruits that we regularly eat today, there are now several cultivars of this exotic fruit that have been raised by growers for more flavour. Another name encountered for persimmons is 'Sharon fruit'. This is a trade name for a type bred and cultivated commercially on the Sharon Plain in Israel, between the Samarian Hills to the east and the Mediterranean Sea to the west. These non-astringent fruit have no core, are seedless and very sweet.

Other species of *Diospyros* include *Diospyros lotus*, the date plum or Caucasian persimmon. This is the most common persimmon planted in gardens and bears the smallest fruits of the group – its common name derives from their date- or plum-like taste. It has a wide natural range and is native from the Caucasus Mountains to China and South Korea. *Diospyros virginiana*, the American or common persimmon grows naturally in the southeastern states of the USA up to Connecticut. It makes a tree to 20 metres (66 feet) high with deeply fissured bark resembling the skin of an alligator. The small round fruits of this persimmon are yellow, with slightly red cheeks, and are rich in vitamin C, making them popular in fruit pies; early European explorers described them as looking like a medlar (*Mespilus germanica*). The fruit, wood and bark were all used by Native Americans, and the dried seeds were made into buttons for the uniforms of soldiers in the American Civil War.

Sago palm

Metroxylon sagu

Not to be confused with the houseplant often called the sago palm, which is actually a cycad (*Cycas revoluta*), or sago pudding, which is usually tapioca made from cassava (*Manihot esculenta*), the true sago palm, *Metroxylon sagu*, provides a staple part of the diet of many people across Malaysia and Indonesia. It grows wild and is also cultivated in several countries in the region and can be found over vast areas of Papua New Guinea, which is known to be a particularly valuable centre of diversity for the species.

The sago palm grows to about 10 metres (33 feet) high, and is topped by a crown of leaves each around 5 metres (16 feet) long. During its lifetime the palm stores starch, which it makes by photosynthesis in its leaves, within each pithy stem or trunk. These can grow to about 75 centimetres (30 inches) thick, although they are larger at the base. In this way the tree builds up enough energy reserves to put out a vast flowering structure and then to set fruit. It does this only once, after which it will die. Such species are often referred to as 'suicide plants', but technically this is a hapaxanthic life cycle. In the wild the sago palm generally only lives for around twelve years, and if allowed to flower and be fertilized it will produce beautiful large spherical fruits. These turn from green to a bronzed brown when dry, with overlapping scales that make them look like botanical pangolins.

Humans take advantage of the plant's energy reserves by cutting it down just before it flowers, or as it begins to flower, in this way harvesting the maximum amount of energy stored in the stem. Traditionally, the starch is scraped out of the trunk by hand, sieved, kneaded and washed before being dried and turned into a flour. Today, fully mechanized factories can carry out these processes, saving time and a lot of effort. The resulting flour is mixed to make a grey, gluey paste, which is eaten with fish or vegetables, or it can be turned into cakes that can

Sago palms grow to around 10 metres (33 feet) high in a relatively short time, providing a valuable food source in their pithy stems. This painting by Marianne North shows them soaring over banana trees in Java.

METROXYLON Rumphii.

OPPOSITE The sago palm will only flower once in its lifetime. It stores starch in its stem, grows a vast flowering structure, sets and disperses its fruit and then dies.

BELOW Naturalist Alfred Russel Wallace observed the laborious process required to create sago flour. The starchy pulp of the palm must be scraped out of the trunk, then sieved, kneaded and washed.

be stored and easily transported. A favourite ingredient, it is used in puddings and jellies, as well as noodles, dumplings and soups. Sago 'pearls' – dried balls of sago cooked in coconut milk and rolled in palm sugar – are a popular food in Sarawak. Sago is thought to be one of the oldest plants to have been exploited and used as a food by people in the tropics. It is a particularly useful plant to grow in wet swampy ground where other crops will not succeed.

The remarkable British Victorian naturalist Alfred Russel Wallace, who independently came up with the theory of evolution by natural selection as well as Charles Darwin, spent many years in Southeast Asia. He travelled widely, collecting thousands of specimens and artifacts, and wrote several books including *The Malay Archipelago,* published in 1869. When in the Moluccan Archipelago off west New Guinea he collected sago cakes and wrote: 'The hot cakes are very nice with butter, and when made with the addition of a little sugar and grated cocoa-nut are quite a delicacy.... When not wanted for immediate use they are dried in the sun for several days.... They will then keep for years; they are very hard, and very rough and dry.'

Sure enough, in the extensive economic botany collections at the Royal Botanic Gardens, Kew, is a stack of rock-hard sago cakes that Wallace collected and sent to Kew in 1858. Although slightly discoloured and looking rather unappetizing, they are proof of the long storage life of this product. They also offer a link across the years, as in many places in New Guinea today it is still possible to see the same style of sago.

Pomegranate

Punica granatum

Early herbals include some curious entries on the pomegranate, mainly drawn from the writings of the first-century AD Greek physician Pedanius Dioscorides. The herbals' authors describe the fruit's pleasant juice and claim it is good for the blood, for the liver, for bleary eyes and heartburn, among other ailments. In Rembert Dodoens' *New Herball* of 1578 he states: 'The juice of the Pomegranate is very good for the stomacke, comforting the same when it is weake and feeble, and cooling when it is to hoate or burning.' Three types of the fruits were documented, and the line drawings on the ageing pages are surprisingly accurate, suggesting a close familiarity with both tree and fruit.

'The juice of the Pomegranate is very good for the stomacke, comforting the same when it is weake and feeble.'

Knowledge and appreciation of the juicy glistening garnet-coloured fleshy seeds of the pomegranate date back much further than these old tomes of herbal wisdom, and even beyond the classical Greeks. Thought to be native to the lands of Iran and northern Turkey, the pomegranate has been cultivated since antiquity around the Mediterranean. Its fruits and flowers have been popular for thousands of years, with references found in ancient Egypt and Mesopotamia. In his tomb, an ancient Egyptian architect and official from the early part of the 18th Dynasty (over 3,000 years ago) called Ineni listed more than 350 plants he grew in his garden at Thebes, among which were five pomegranate trees. One of the pharaohs for whom Ineni worked, Tuthmosis III, also loved gardens. Wall reliefs carved in a room now known as the Botanical Garden in the temple of Amun at Karnak depict all the exotic plants Tuthmosis had collected on his campaigns in Asia Minor, including the easily recognized pomegranate.

Among the valuable cargo discovered on an ancient shipwreck at Uluburun off southern Turkey, dating to 1306 BC, were storage jars

The pomegranate features in the *Tacuinum Sanitatis*, a beautifully illustrated eleventh-century treatise on health. Later, Rembert Dodoens recommended pomegranate to protect against 'wambling of the stomacke'.

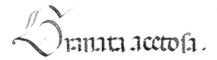
Granata acetosa.

Granata acetosa. ɔplo.frī.   Acetō q̃ sunt multe sucositatis. iuuant tir epī cū. ɔfer.
nocumtum nocent pectori. Remō nocumti cum calce melito. Quīo gn̄ant·chimum·mo
dicum. Nag ɔueniūt calis.iuuenibz,ctate. calɔ regioni.·

OPPOSITE The distinctive fruits of the pomegranate are said to symbolize prosperity, virtue and fertility, perhaps because of the large number of seeds they contain.

BELOW The pomegranate appears in beautiful texts from around the world, including this nineteenth-century Japanese illustrated manual, *Honzō zufu*, by Iwasaki Tsunemasa.

full of pomegranates. Other goods carried on the ship included ebony, terebinth (a fragrant resin), elephant tusks and copper ingots, suggesting that pomegranates were likewise regarded as luxury items. Pomegranates are also frequently represented in ancient art in the form of pottery vessels, jewelry and decorative objects made of glass, ivory and precious metals, and in scrolls, mosaics and coins, further attesting to the status of this fruit.

Since these early times, the pomegranate has been entwined with the cultures of the Mediterranean and Near East in food and drinks and as a medicine. It has also been important in religions including Zoroastrianism, Judaism, Christianity and Islam, usually as a symbol of prosperity, virtue and fertility (perhaps because of its many seeds). This symbolic role has continued into more recent centuries, and the fruit has been incorporated in works by renowned artists such as Sandro Botticelli in his *Madonna of the Magnificat* (1483) and Dante Gabriel Rossetti in *Proserpine* (1874).

Rossetti's painting of course refers to the Greek myth of Persephone and Hades, the god of the underworld. According to this myth, Hades fell in love with Persephone, who was the daughter of Zeus and Demeter, the goddess of nature and harvests, and kidnapped her, taking

her to his realm in the underworld. Demeter begged Zeus to make Hades return her daughter. Hades finally agreed to release Persephone, but only after he had tricked her into eating some pomegranate seeds. Because she had eaten in the underworld she had to return there for several months each year, a time when no crops would grow on the earth, thus explaining the seasons.

Catherine of Aragon, who married Henry VIII of England in 1509, incorporated the pomegranate in her badge or emblem, as a symbol of fertility and regeneration. It is thought that the first tree was brought to England later, around 1610, by the influential plant collector John Tradescant the Elder, who acquired it in Paris. It was introduced to America in the eighteenth century, perhaps first by Spanish settlers, and Thomas Jefferson planted pomegranates in his 'fruitery' at Monticello in 1771.

Today, the original species and over 500 cultivars are grown around the world, from China, Japan, Southeast Asia and Afghanistan to California and Arizona in the USA, thriving where the summers are warm and dry. A small tree, it can reach between 6 and 10 metres (20–33 feet) high if left to grow naturally, but is usually cultivated in orchards of pruned, multi-stemmed trees. Its showy red to red-orange flowers bloom from May to autumn, held at the tips of stems singly or in small clusters, inviting a variety of insect pollinators. The Latin name can be translated as 'apple with many seeds', and this is a good description of this fruit or berry. The number of seeds in a pomegranate can vary from 200 to 1,400, each one surrounded by the sweet fleshy coat, or aril, and all held packed together by a soft white membrane enclosed in a leathery skin.

Whole seeds are used in a variety of dishes, and a refreshing juice is squeezed from the fruits, which can be turned into the cordial grenadine or thickened into pomegranate molasses. Down the ages, bark, fruit and flowers have also been used in a wide range of medicinal remedies as well as love potions. Today we know the fruits (especially in some cultivars such as 'Wonderful') to be very high in polyphenol antioxidants, which may lower the risk of heart disease and cancers. Much research is currently being conducted into the health benefits of pomegranate, including its effects on diabetes and high blood pressure, and it is now regarded as a 'superfood'. The pomegranate has been highly esteemed for millennia, and people continue to devise new ways to eat and enjoy this unusual fruit, which has a history and appearance like no other.

Woods Salcombe
Aug 10. 1895.

Pinus
Pinea.
The stone or umbrella Pin

Stone pine

Pinus pinea

The dark silhouette of the stone pine against azure blue skies is a very familiar sight in the Mediterranean coastal areas of France, Italy, Greece and Spain. The tree's overall shape and habit, with long, spreading branches radiating outwards to form a canopy atop a bare slender trunk up to 25 metres (82 feet) tall, resembles a large parasol, hence its other common names: the umbrella or parasol pine. Stone pines often lean at strange, picturesque angles, and adding to their distinctive character is the deeply fissured, reddish-brown and plate-like bark, which is also fire resistant. Although it is often regarded as an iconic landscape element of Italian Renaissance gardens and the symbol of Rome today, the stone pine has a naturally wide range of distribution. It occurs throughout the forests and maquis scrublands of southern Europe, Israel, Lebanon and Syria, where it grows with the Aleppo pine (*Pinus halepensis*), the evergreen holm oak (*Quercus ilex*) and the cork oak (*Quercus suber*; see p. 28).

The edible seeds of this tree have been used for thousands of years as a culinary item, and the Romans seem to have been particularly fond of them.

One of the famous archaeological sites in Italy is the Appian Way or *Via Appia*, an early Roman military road connecting Rome to Brindisi in the southeast of the country. Named after Appius Claudius Caecus, a Roman censor who built the first section in 312 BC during the Samnite Wars, it is now a popular tourist attraction, lined with the characteristic stone pines. The trees were probably first planted so that their umbrella-shaped canopies would provide much-needed shade for the marching legions of the Roman army during the heat of the day – a function their modern replacements perform for visiting tourists today.

Of all the 100 species of pine, the stone pine cone takes the longest time to mature, at more than thirty-six months. When ripe, the heavy

The flexible needle-like leaves of the stone pine, 10–20 centimetres (4–8 inches) long, are held in bundles. The seed-bearing cones are 8–15 centimetres (3–6 inches) long, and take thirty-six months to mature, longer than any other pine.

cone releases stone-hard shells containing the edible seed or pine nut, also known as piñon or pinoli. It is this hard shell surrounding the seed that is said to be the origin of the more popular common name 'stone pine'. One thousand seeds weigh approximately 718 grams (25 ounces), and even with the small wing attached to each one, they are too heavy to be effectively dispersed by the wind, so the tree relies on animals, in particular the Iberian magpie, rodents and, more recently, humans, to disperse them. The edible seeds have been used for thousands of years as a culinary item, and the Romans seem to have been particularly fond of them – the hard shells have even been found in excavated refuse heaps in Roman encampments in the cold, far-off, northern province of Britain, having been sent over as food for the armies, who considered them a delicacy.

Today millions of kilograms of pine nuts are harvested commercially each year, using long, hooked poles to collect the ripe, unopened cones, which are then heated to release the seeds. They are high in protein and thiamine (vitamin B) and widely used in French and Italian culinary dishes, being one of the main ingredients of pesto sauce, along with basil, parmesan or pecorino cheese, garlic, salt and olive oil. In Catalonia, another popular recipe using pine nuts, which dates back to the eighteenth century, is a dessert called 'Panellets' – small round cakes made from a ball of marzipan covered in pine nuts. Despite the import of cheaper pine nuts from China, harvested from the Chinese white pine (*Pinus armandii*), and from South Korea from the Korean pine (*Pinus koraiensis*), kernels from the stone pine are reputed to have the best flavour and without the bitter taste. Resin is also tapped from the trunk and used for varnish and as a source of rosin for violin bows and ballet shoes.

Although essentially Mediterranean, the stone pine is now grown elsewhere in the world where the climate is suitable. It is believed that it was introduced to Britain in the sixteenth century and has been planted as an ornamental tree since the eighteenth century. A curious early planting is found in the arboretum at the Royal Botanic Gardens, Kew. Planted in 1846 by Sir William Hooker, the tree was kept in a container for many years and had become bonsai-like, forming several large branches growing from the trunk at only 1 metre (3 feet) above the ground. It has now matured to create a unique shape, very unlike the tall, bare trunks and stately growth habit typical in its native home.

ABOVE Ripe, unopened cones of the stone pine are collected and then heated to release the seeds, known as pine nuts. These are surrounded by a hard shell, thought to be the origin of the usual common name for the tree.

OPPOSITE The general form and habit of the stone pine resembles a large, broad parasol, and the tree is also known as the umbrella or parasol pine.

Pinus Pinea Lin.

Fish poison tree, *Barringtonia asiatica*

HEALERS AND KILLERS

Plants produce the most astonishing array of active compounds in order to help themselves stay healthy and also to prevent them from being eaten by predators. We, mostly by trial and error, have come to learn how to use the most potent pieces of a plant's armoury – whether for good or ill.

At least 28,187 plant species are currently recorded as being of medicinal use, according to the 2017 *State of the World's Plants* report by the Royal Botanic Gardens, Kew, and in many areas of the world these plants can be the main sources of medicine. However, plants should not be regarded as simply 'folk-medicine', only to be used when nothing else is available. They are powerful healers and new pharmaceutical drugs are being developed from the active compounds found within them every year. For instance, the ginkgo has a long history of use in traditional Chinese medicine for a great range of ailments, but is now also being investigated for applications in modern medicine.

One of the most famous drugs in history derived from a tree is surely quinine (from *Cinchona*), which has saved countless millions of lives over the past 200 years as a preventative and cure for malaria, and has allowed people to live in areas where they would otherwise be at risk. More recently, the drug Paclitaxel has been developed as an anti-cancer drug from the bark and leaves of yew trees (see p. 58). Trees do not have to have such dramatic medicinal effects to be useful, however, for there are plenty that are beneficial in other ways in our everyday lives. The neem tree has such a wide range of anti-bacterial properties that in India it is known as the 'village pharmacy', and is even used in toothpaste. And the tea tree from Australia has become a staple over-the-counter antiseptic for our medicine cabinets.

But as much as plants can heal us they can also harm, and sometimes it is even the same species that can do this, as it can be purely the dose which can kill or cure. There are many species, however, that simply should not be messed with. Strychnine can cause one of the most horrific deaths imaginable, but until relatively recently was kept in people's homes as a rat poison. The 'apples' of the highly toxic manchineel are most definitely to be avoided, and even the tree's milky sap can cause painful blistering. The fish poison tree lives up to its name, though in low doses is also used in traditional medicine to treat various conditions. Less deadly, but certainly still causing discomfort, the leaves of the headache tree release a volatile oil that can cause headaches, but conversely are also thought to be able to cure them.

Maidenhair tree

Ginkgo biloba

Ginkgo biloba is a long-lived native tree of China, with some individuals reputed to be over two and a half thousand years old. But the history of this tree stretches much further back in time. The maidenhair tree or ginkgo is a living fossil. During the time dinosaurs roamed the Earth, some 200 million to 175 million years ago, ginkgos were widespread across the northern hemisphere. Fossil leaves of ginkgo, albeit an ancient species, *Ginkgo huttoni*, are found preserved in rocks.

The name *Ginkgo* has an interesting history in itself and is thought to be a mistake in the transliteration of the Japanese name for the tree, 'ginkyo', meaning 'silver apricot'. It appeared in the familiar form in the first description of the tree in the West, by the German botanist Engelbert Kaempfer, who encountered a ginkgo in the garden of a Japanese temple when he was in Nagasaki in 1690–92. The mistake was then perpetuated when the tree was scientifically described by Carl Linnaeus in 1771; he added the specific epithet *biloba*, which derives from the Latin *bis*, 'two', and *loba*, 'lobed', referring to the unique, bi-lobed shape of the leaves. Ginkgo is also used as the common name, while another name is maidenhair tree, which alludes to the resemblance of the leaves to the maidenhair fern (*Adiantum capillus-veneris*). The unmistakable shape of the leaves has led to the tree's name in Chinese, 'I-cho', meaning duck's foot, and it is also called the 'godfather tree' because a tree planted by one generation will begin to bear fruit one or two generations later.

Ginkgos are beautiful trees that grow to 20 to 30 metres (66 to 100 feet) tall, with a very distinctive sparsely branching form. On a sunny autumnal day they bring the landscape alive as their leaves

> *The unmistakable shape of the leaves has led to the tree's name in Chinese, 'I-cho', meaning duck's foot.*

The maidenhair tree was scientifically described as *Ginkgo biloba* by Carl Linnaeus in 1771. He gave it the species name *biloba*, Latin for 'two' and, 'lobed', referring to the shape of the leaves. Ripe fruits produced on female trees have an unpleasant smell, but the seed kernels are regarded as a delicacy. This image was commissioned from a Chinese artist by Robert Fortune when he was in China in the mid-nineteenth century.

HEALERS AND KILLERS

HEALERS AND KILLERS

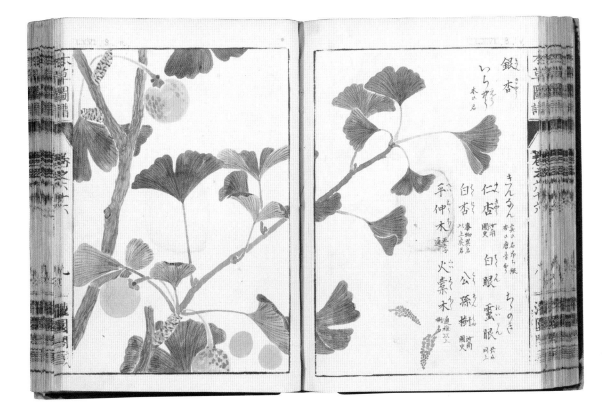

OPPOSITE One of the early maidenhair trees introduced to Europe was planted in the new arboretum at the Royal Botanic Gardens, Kew, in 1762. It is a male tree and produces yellow cone-like flowers. In autumn the leaves turn from green to a rich golden yellow before falling.

ABOVE The ginkgo is native to China, where it has long been grown. It was first introduced to Europe from China via Japan around 1730. This Japanese illustration is from *Honzō zufu,* a botanical work that eventually consisted of 93 volumes.

change colour from green to the most brilliant golden yellow before falling. Once classified as a conifer and placed in the same family as the yew tree (Taxaceae), the ginkgo is now separated from the conifers and placed in an order all of its own, the Ginkgoales, of which there is only this single species. It is one of the trees that is dioecious, meaning there are separate male and female trees. The males produce tiny 'cones' containing the pollen, while the females bear a hard-shelled seed covered in a fleshy outer skin, which when ripe is a light-yellow colour and similar in appearance and size to an apricot. Unfortunately, as well as being beautiful, the ginkgo also has a reputation for being one of the smelliest trees in the temperate world. The flesh around the seed contains butyric acid and when ripe gives off the smell of vomit. It is thought that this odour, also like rotting flesh, attracts nocturnal animals which eat the fruit and then distribute the seeds.

Despite the smell, the seed kernels are prized as a delicacy and are used in many dishes in eastern cuisine, and are also thought to have aphrodisiac qualities. There is also a long history of use of the leaves and seeds in traditional Chinese medicine for the treatment of ailments including digestive problems, asthma and lung complaints.

More recently research has shown that the antioxidants in the leaves can increase blood flow, which may have applications in the treatment of diseases relating to poor blood circulation. It is also claimed that extracts may improve memory and concentration and help in dementia. However, it should be noted that there is no clinical evidence for the effectiveness of ginkgo in the treatment of the numerous ailments it is said to alleviate, and further scientific research is needed to ascertain the efficacy of ginkgo medicinally.

The largest and possibly the oldest ginkgo tree in the world, estimated to be between 4,000 and 4,500 years old, is the Li Jiawan, 'Grand Ginkgo King', in Guizhou province, China. A male tree, it is about 30 metres (98 feet) tall with a completely hollow trunk that measures 4.6 metres (15 feet) in diameter, or 15.6 metres (51 feet) in circumference. There are many other ancient specimens around the world, particularly associated with temples in China, Korea and Japan. One magnificent tree growing in a small town called Lengji in Sichuan province even has a temple inside it. In 2005 it measured 30 metres (98 feet) tall with a girth of 12.4 metres (40 feet). It is said to have been planted when the politician and writer Zhuge Liang went on one of his military expeditions during the Three Kingdoms period, which would make the tree more than 1,700 years old.

The ginkgo was first introduced into Europe from China via Japan around 1730, and one of the early introductions was planted in the new arboretum at the Royal Botanic Gardens, Kew, in 1762. It was trained like a fruit tree against the wall of the Great Stove, a glasshouse built for Princess Augusta in 1761, probably to give it some winter protection, though the tree is hardy. It may well be one of the first to be planted in Britain and is now one of the 'Old Lions of Kew'.

Ginkgo trees make the perfect urban tree – 17 per cent of Japan's street trees are ginkgos – because of their resistance to pollution and disease. Today cultivated forms of known male trees such as 'Princeton Sentry' and 'Autumn Gold' are planted for their horticultural merits and tolerance of urban pollution, and to avoid the problem of the foul-smelling fruits.

Fish poison tree

Barringtonia asiatica

Beautiful, useful, fragrant and poisonous: this species has an impressive list of attributes to go with an imposing and appropriate English common name. A tall tropical tree from Asia, Madagascar and the Pacific islands, *Barringtonia asiatica* has found its niche growing along coastal and river shorelines and in wetlands among mangrove species, where it thrives with its roots dangling in the water. Reaching up to 20 metres (65 feet) high, with a lush billowing canopy, a mature specimen of this tree can be a spectacular sight, especially when in full bloom.

The beautiful pom-pom-like flowers of the fish poison tree develop into four-cornered fruits. These contain the saponin-rich seeds that are used to stun fish.

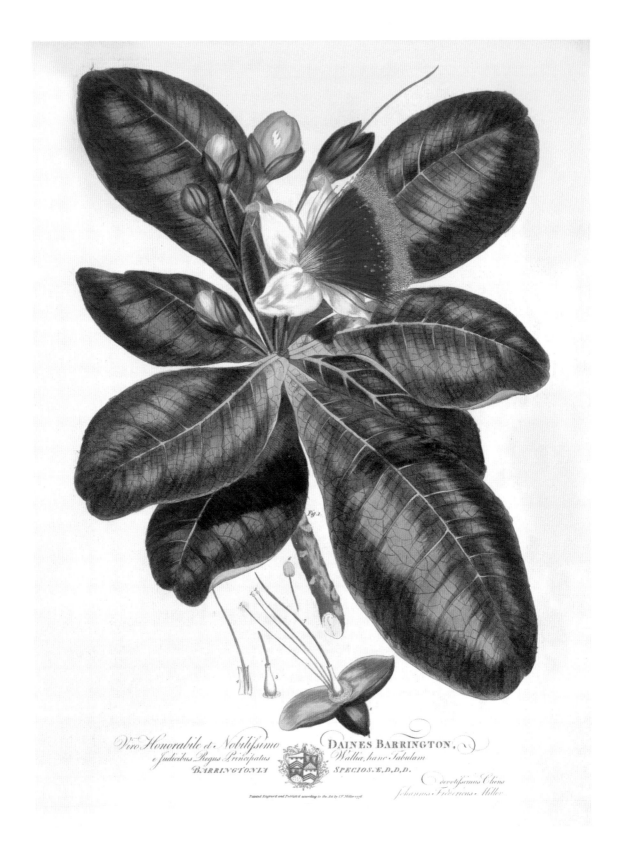

Viro Honorabile et Nobilissimo   DAINES BARRINGTON,
e Judicibus Regis Principatus   Walliæ, hanc Tabulam
B. ARRINGTONIA   SPECIOS. E. D, D, D.

devotissimus Cliens
Johannes Fredericus Miller

Painted Engraved and Publish'd according to the Act by I.F. Miller 1776.

While *Barringtonia asiatica* may be poisonous, it is also an extremely attractive tree, with large, glossy leaves and flowers with a striking cluster of stamens. The small white petals at the base of the flowers are easily overlooked.

The flowers resemble large pom-poms, an effect caused by a mass of showy pink-tipped stamens bursting from between four small rounded white petals, which are easily overlooked. At dusk, the blooms give out a heady, almost sickly, scent, and the pale colours and fragrance attract nocturnal pollinators including bats, flying foxes and moths. After pollination, four-cornered fibrous fruits develop looking like small lanterns. These are also known as box fruits and contain one to two oblong seeds. The fruits are spongy in texture and have the advantage of a waterproof skin, so that when they drop into water they can float away to new suitable areas with the right conditions for germination. Generally considered to be able to stay afloat for many months, the fruits may even survive for up to fifteen years according to one study. The ability to spread in this way as 'drift fruits' has resulted in this tree being widely distributed, and it has in fact become invasive in some coastal areas.

Generally considered to be able to stay afloat for many months, the fruits may even survive for up to fifteen years according to one study.

The fish poison tree contains a bitter and toxic saponin (a phytochemical thought to deter animal foragers by making the plant unpalatable), which is particularly concentrated in its seeds. People pound the seeds into a fine pulp, which is then thrown into streams. Because the toxic saponin readily dissolves in water it can stun fish, making them easy to gather up. This is one of many tropical plants used as a fish poison (or piscicide) in this way.

As with many potentially poisonous species, it is the dose of a plant chemical that can make it helpful or harmful. In low doses, the seeds, bark and leaves of this tree are used as a traditional medicine for treating coughs and lung problems, as well as for intestinal worms and other parasites. In the Philippines the leaves are heated and applied externally to relieve stomach ache, headaches and joint aches, while in the Bismarck Archipelago near New Guinea it is reported that the fresh fruits are applied to skin sores. In Indochina the fruits are boiled to remove the saponins and then eaten as a vegetable. The obvious charms of the flowers and the large, glossy leaves of *Barringtonia asiatica* have also led to it being planted as an ornamental shade tree in India and Singapore, where it is greatly admired.

Headache tree

Umbellularia californica

More of an inconvenience than an actual threat to life, the headache tree is certainly still a species to avoid sitting under. It was highly valued by the Native American peoples of California, who used the leaves and fruits for various purposes, including as a poultice to ease headaches. Conversely, however, a volatile oil released by the leaves can actually have the opposite effect.

Found in the coastal woodlands and redwood forests of California and into Oregon, this large evergreen hardwood tree can grow up to 30 metres (100 feet) tall. Its pungently aromatic leaves contain a high percentage of umbellone, which when breathed in can cause severe headaches and migraines in those who spend time around the trees. The active compound can also produce irritation to the eyes and nose. Despite this disadvantage, this fine tree is often planted as a flowering ornamental specimen or hedge in gardens. In one reported case an Italian gardener working in California had suffered cluster headaches for twenty years without knowing why, until it was discovered that pruning this tree was the cause.

> *Its pungently aromatic leaves contain a high percentage of umbellone, which when breathed in can cause severe headaches and migraines.*

The species has many common names including Californian laurel, pepperwood and spice tree, and it can be easily confused with bay laurel (*Laurus nobilis*) as they are in the same family (Lauraceae). The thick green leaves are lance-shaped, and the small yellowish flowers develop into egg-shaped glossy green berries. These mature to a deep purple colour and have large brown seeds or 'nuts' inside. Although parts of the tree, mainly the leaves and 'nuts', are edible, they are generally avoided except by the more adventurous forager.

The famous Scottish plant hunter David Douglas, who travelled widely in America's northwest, came across the tree in Oregon in 1826

Also known as the Californian laurel, the headache tree can easily be confused with the bay laurel as they look similar and are related. It is often planted in gardens, but can have unpleasant effects.

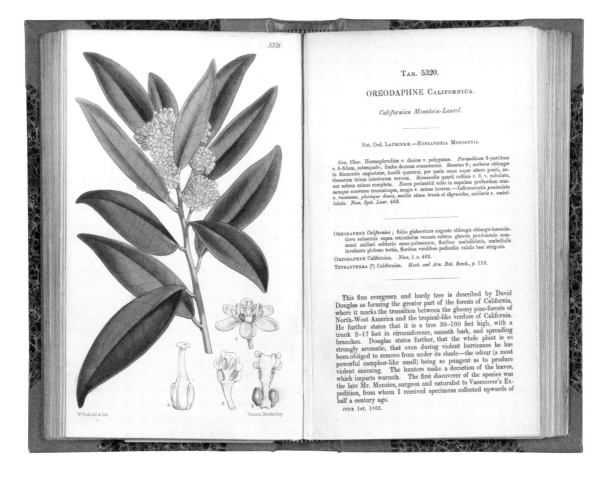

and collected its seeds. He described it as a 'beautiful evergreen tree' and said the local people made a beverage from its bark, though he also mentioned that the powerful scent could cause sneezing. Douglas shipped the first specimens back to Britain in 1829, where they were planted at the Royal Botanic Gardens, Kew, and in several country estate gardens. It is recorded in *The Hillier Manual of Trees and Shrubs* that 'the "old school" of gardeners indulged in extravagant stories of the prostrate Dowager overcome by the powerful aroma' of this unusual tree.

Manchineel

Hippomane mancinella

A member of the spurge family (Euphorbiaceae), this species actually holds the record as the world's most dangerous tree. The milky sap of the manchineel, which drips from any wounds in its trunk or branches, as with other spurges, contains strong irritants. It is so caustic that on contact with the skin, the sap will immediately cause blistering and burns, and can produce temporary blindness if it gets in the eyes. Even standing under this tree in the rain is dangerous, as drops contaminated by the sap can have the same effects.

Native to the tropical areas of southern North America (including Florida), the Caribbean, Central America and northern South America, this evergreen tree grows up to 15 metres (50 feet) tall. It is found along beaches and coastlines, where its roots help prevent erosion. The fruits resemble small green apples, but they are also highly toxic and the tree has many sinister common names including the Spanish *arbol de la muerte* or *manzanilla de la muerte* – tree or apple of death. Said to taste quite sweet, the fruit's flesh if eaten soon results in severe burning and ulceration of the mouth and throat, leading to excruciating pain. As all parts of the manchineel are toxic, local people will sometimes mark the trunk of a tree with a red X or a sign to warn of its presence. The wood is used, with care, in the making of furniture, but even burning it is dangerous as the smoke from the fire can still give rise to serious eye problems.

Encounters with this species are mentioned by several famous explorers. The eighteenth-century naturalist Mark Catesby recorded the agonies he suffered after the juice of the tree got into his eyes, and that he was 'two days totally deprived of sight'. Manchineel's notorious reputation has even spread into literature – references are found in *Madame Bovary* and *The Swiss Family Robinson*, among others, while it also appears in operas, including Giacomo Meyerbeer's *L'Africaine*, where it is chosen as a means of suicide by the heroine Sélika.

OPPOSITE People have learnt the hard way to avoid the sap and tempting fruits of the manchineel, which are highly toxic. The tree has evolved this severe deterrent to rebuff herbivores, while its fruits are thought to be designed to drop on to the beach and be carried away by the tides to a new location where they can germinate and grow.

RIGHT Many explorers including Captain Cook came across the manchineel tree and saw its effects first-hand. The Spanish explorer Juan Ponce de Leon who died in a skirmish with native people in southwest Florida in 1521 is said to have been killed by an arrow dipped in manchineel.

Strychnine

Strychnos nux-vomica

———

Strychnine, notorious in both fact and fiction as a fast-acting and deadly poison, is a natural alkaloid found in the seeds of a tree of the same common name. If ingested or absorbed it is often fatal, usually proving deadly within three hours. Plants have evolved such alkaloids, and can concentrate them in certain parts such as seeds, bark and roots, to protect themselves and to deter unwelcome predators from eating them. Strychnine certainly takes no chances.

If ingested or absorbed it is often fatal, usually proving deadly within three hours.

The main species from which strychnine is derived, *Strychnos nux-vomica*, is native to India and Southeast Asia. Also known as the poison nut, this medium-sized tree, usually 10–13 metres (33–43 feet) tall, develops a cluster of small green-white flowers (technically a cyme) that emit a disagreeable smell, an indication of the much more disagreeable seeds that will follow. It is these flat, disc-like seeds, which develop inside the reddish-yellow, apple-sized fruits, that are the source of the dangerous alkaloid, alongside another called brucine. Around five seeds are enclosed within a white sticky pulp inside the fruits, each coated with soft hair. The ashen-grey seeds can easily be removed and cleaned, and are then ground into a powder. Strychnine powder has been used as both a 'medicine' (usually as a stimulant) and a poison – depending on the dose. It has a long history in Indian and Arabic medicine, though it should be noted that strychnine has no known medical value.

Trade with India in the sixteenth century brought strychnine to the attention of the West, and it was soon being employed as a rat poison on ships and in houses. It can cause a particularly gruesome death as it binds to the glycine receptor of the nerve synapses, taking the brakes off and firing the nerves at the smallest stimulus. This leads to horrific muscular convulsions, with victims sometimes arching their backs to

The reddish-yellow fruits of strychnine contain a white pulp that holds several grey seeds. These are the source of the notoriously deadly poison.

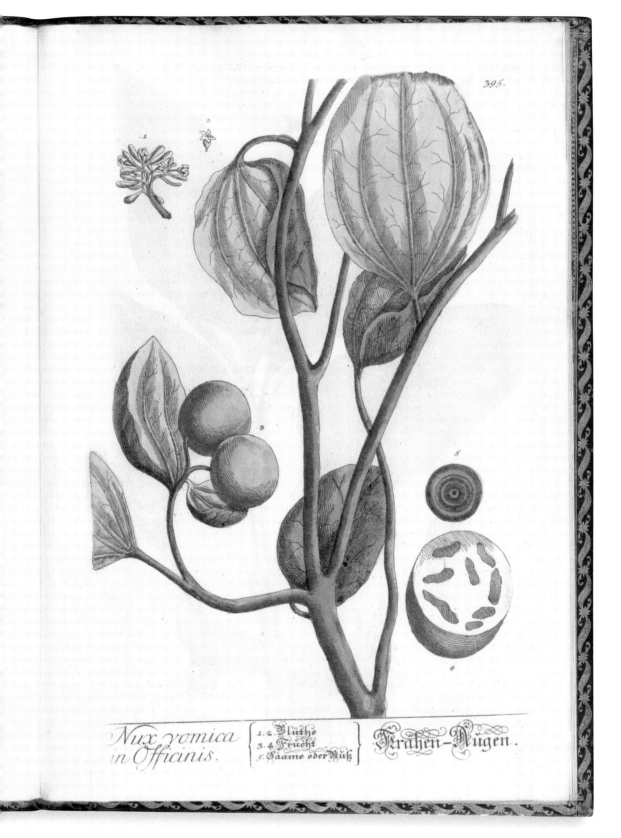

395.

Nux vomica in Officinis.

[1. 2. Blüthe
3. 4. Frucht
5. Saame oder Nuß]

Krahen-Augen.

The seeds of strychnine are distinctively flat and disc-like, and can be ground up to make a fine powder. This was once commonly available and used in homes as a poison, but is now a banned substance in many countries.

the extent that they balance on the backs of their heads and their heels. Death usually follows by asphyxia as the nerves controlling breathing fail to function properly. Throughout much of this, the victim is acutely aware of what is happening, making it a particularly cruel and dramatic way to die.

Strychnine is very bitter and not particularly soluble in water, so fortunately it is not that easy to ingest by mistake, although this did happen on numerous occasions when people bought and kept bottles of the poison in their homes. Agatha Christie used it several times in her novels, including her first, *The Mysterious Affair at Styles* (published in 1921), which introduced Hercule Poirot. In her description of the murder of Emily Inglethorp, Christie revealed her extensive chemical knowledge. Mrs Inglethorp had been prescribed a strychnine tonic to invigorate her, but foul play leads to an overdose.

Poisoning by strychnine is not just a feature of fiction, however, as it has also played a major role in numerous real-life murder cases. One of the most notorious was that of Dr Thomas Neill Cream, who when living in Canada killed at least two women in the 1870s as well as a man in 1881, for which he went to prison. Once released, he travelled to Britain and proceeded to poison several prostitutes in London using strychnine. He became known as the Lambeth poisoner and was eventually convicted and hanged in 1892.

Products containing strychnine are now highly restricted in many countries and it has been banned in the United Kingdom since 2006; it is illegal to purchase it in any form. There is no specific antidote for strychnine poisoning, although treatment with diazepam and muscle relaxants can help. A victim of poisoning needs medical treatment as early as possible to have any chance of survival.

Other members of the same genus as *Strychnos nux-vomica* also produce deadly poisons, including *S. ignatii*, or the St Ignatius bean, from the Philippines, and *S. toxifera* of South America, from which curare is derived.

Neem

Azadirachta indica

This widespread species is possibly one of the most useful trees in the world, and it is probably used daily by people in India, where some call it the 'village pharmacy'. If you're going to have one tree near your house – if possible, this should be it.

Native to Assam, Bangladesh, Cambodia, East Himalaya, Myanmar and Thailand, the neem tree (*Azadirachta indica*) has been introduced across India and other parts of Asia, as well as in Africa and even the Caribbean. It is a member of the mahogany family (Meliaceae), and is a fast-growing evergreen tree, generally 15–20 metres (50–75 feet) tall, thriving in tropical and sub-tropical regions. With its dense crown of delicate pinnate leaves it can cast a pleasing shade, and its drooping panicles of pretty white flowers give off a beautiful fragrance. These are followed by small, greenish-yellow fruits. In some areas the neem tree grows so well it may even be considered a weed, but its uses far outweigh any inconvenience.

The young leaves and flowers can be cooked and eaten, although they are said to be bitter. But it is as a source of medicinal products that the neem tree is so valuable and that has made it a feature of traditional Chinese, Ayurvedic and Unani medicines for thousands of years. The leaves and seeds contain hundreds of active ingredients which act as antioxidants and are thought to impede the growth of bacteria. Neem is also believed to have anti-fungal and anti-inflammatory properties and is even being investigated for its potential to inhibit cancerous

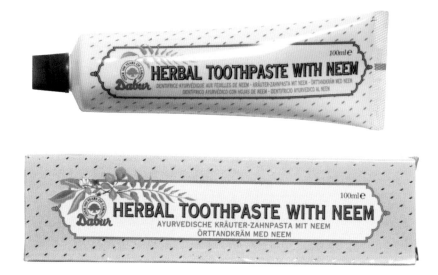

tumours. In traditional medicine, products made from neem leaves are
used for skin conditions, while oil pressed from the seeds is believed to
improve liver health and to 'cleanse the blood'. In countries where the
tree grows neem oil is also used in many cosmetics such as soaps, shampoos and face creams. People use neem twigs as a natural toothbrush,
and you can find neem toothpaste and mouth-rinses on sale.

But the myriad benefits of the versatile neem do not even end there.
The tree plays an important role in protecting homes and crops from
insect pests. The leaves are dried and placed in wardrobes and kitchen
cupboards to repel insects, and can also be burned to ward off mosquitoes. A powder made from the seeds and diluted in water is sprayed on
crops as a natural, biodegradable pesticide, deterring insects from feeding and laying eggs and affecting their life cycle. Although it has to be
applied frequently, it is safe, easy to obtain and is an environmentally
friendly option that helps protect the yields of many crops – it has even
been found to discourage locusts.

Because of its importance in daily life and great usefulness, the flowers and leaves of this 'wonder tree' are commonly used in traditional
Hindu festivals in India, including the Mări or Mariamman festival in
the state of Tamil Nadu, which celebrates the rain and mother goddess.

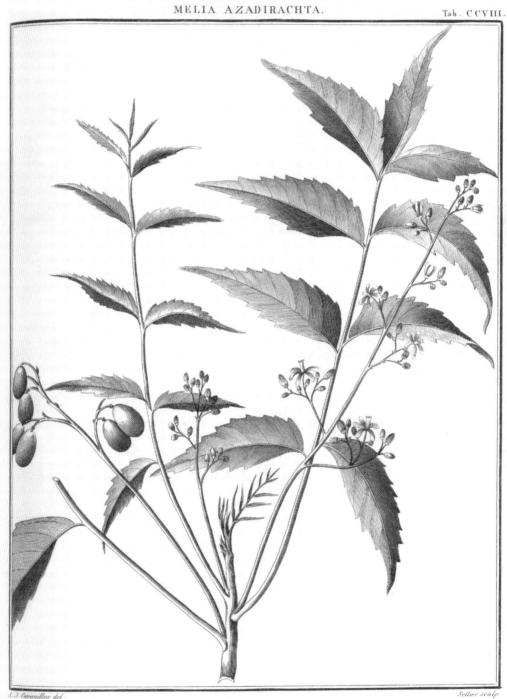

A.J. Cavanilles del.   Sellier sculp.

HEALERS AND KILLERS

Quinine

Cinchona spp.

The tale of the *Cinchona* tree is one of medical discovery, botanical smuggling and the expansion of empire. For almost 400 years this tree has been of vital importance across continents for its life-saving properties. All this is because the bark of *Cinchona* provides quinine – a natural alkaloid that can be taken as a preventative and cure for malaria, one of the world's biggest killers.

There are around twenty-three species of *Cinchona*, which are native to the Andes and can be found in Ecuador, Bolivia and Peru. They all look very similar and hybridize freely, but two species, *C. officinalis* and *C. pubescens*, have been of particular importance in the story of quinine. Although known purely for the usefulness of their bark, these small trees also have pretty pink-purple tubular flowers, which are pleasantly scented and attract the butterflies and other insects that act as pollinators in the trees' native habitats.

First recorded as being useful against fevers by Spanish missionaries in Peru in the 1630s, the reputation of the bark of the 'quinquina' trees for curing malaria quickly spread across Spain and the rest of Europe after being introduced by the Jesuit order. Supplies to Europe were sporadic and highly priced at first, but the efficacy of 'Jesuit's bark' was soon established. Unfortunately, Protestant England initially remained sceptical because of the Catholic associations, but the bark was certainly being used in the country by the end of the seventeenth century.

The botanist and father of taxonomy Carl Linnaeus named the genus *Cinchona* after the beautiful Spanish Countess of Chinchón, who, it was rumoured, was one of the first Europeans to be cured of malaria or 'ague' by the bark of the quinine tree in 1638 (sadly, this part of the story seems not to be true). Linnaeus' misspelling of her name has led to confusion about how to pronounce the genus ever

In the late nineteenth century *Cinchona* plantations were created in countries close to areas where supplies of quinine were most needed to combat malaria.

since. For the following two hundred years 'the bark' was harvested, exported and exploited as the main treatment for combating malaria. Its role in the expansion of European empires across the tropics is undisputed. Where once the fatality rate for Europeans in Asia and Africa was extremely high, limiting their ability to become established, a supply of quinine made a dramatic difference to the health and success of expeditions and colonies.

It was in the early nineteenth century that the active alkaloids that give *Cinchona* bark its potency were discovered. In fact four alkaloids have been found to kill the *Plasmodium* parasite that causes malaria, but because of its strength and effectiveness quinine became the most important. The French chemists Joseph Bienaimé Caventou and Pierre Joseph Pelletier were the first to isolate quinine and analyse it in 1820, and production of pure quinine began soon after.

It was also found that *Cinchona* trees from different regions varied in the quantities of alkaloids they contained. Although influential 'quinologists', including the Englishman John Eliot Howard, began to grow, manufacture and study *Cinchona* to research which trees were the most effective, supplies of the bark and quinine as a product still fell far short of demand. The sources of the trees in South America were fiercely guarded by growers to maintain their monopoly on such a profitable product. This is where the botanical spies and smugglers enter the story, with the aim of obtaining sufficient seeds and plants to start plantations nearer to where the quinine was most desperately needed: a dark deed justified as philanthropy to save countless lives.

In the archives of the Royal Botanic Gardens, Kew, is an intriguing collection of letters of past directors. One, dated 1859, from the botanical explorer Richard Spruce to Kew's first director, Sir William Hooker, describes an attempt to obtain quality seeds in Ecuador. Spruce was the leader of one of three British government-funded expeditions, which also involved two Kew gardeners, to seek out the best *Cinchona* trees. Eventually they brought back 100,000 plants and 637 seeds. In 1860 Kew built a 'Cinchona Forcing House', paid for by the India Office, and by 1861 seedlings had been raised that were destined for plantations and gardens in India's Nilgiri Hills and Ceylon (Sri Lanka). Once the plantations were established, quinine became widely available to both the local and British resident populations.

The next development in the story is how, in order to consume their daily dose of quinine, British colonials added it to a 'tonic water', but as that was still bitter, they mixed it with lemon, sugar and gin, and the G&T was born. The first British patent for quinine in a carbonated tonic water dates to 1858.

With over a thousand items, Kew has one of the largest collections in Europe of *Cinchona* bark and seeds, with some dating back to the early 1700s. Shelves are packed with rows of old glass jars, black boxes and paper packaging with Victorian handwriting, each containing a specimen sent from a different place or time. A large proportion of this originates from John Eliot Howard in the 1860s, as he became a prolific collector and had his own quinine manufacturing business. There is also a sample of seeds dated 1865 which was brought to Kew by a knowledgeable but unlucky collector named Charles Ledger. He had sought out the seeds of the best-quality *Cinchona* trees in Bolivia with the help of a local man named Manuel Incra Mamani. Unfortunately, as Kew already had samples, as well as plants growing in India, it refused to buy Ledger's seeds, so he struck a deal with the Dutch instead. The Dutch then went on to dominate the global quinine market, as the trees grown from Ledger's seeds contained the highest quality quinine, just as he and Mamani had claimed.

Today, although the use of quinine has declined as a result of the development of other effective drugs and the increasing resistance of the *Plasmodium* parasite that causes malaria, it remains extremely important, and has also inspired medical advances in other anti-malarial drugs. *Cinchona* can truly be viewed as a tree that has changed the course of world history and saved countless millions of lives.

Although mainly valued for their medicinal bark, *Cinchona* trees are also beautiful in their own right, producing pleasing pink tubular flowers, which are attractive to pollinators.

Tea tree

Melaleuca alternifolia

In its native Australia, the tea tree has long been well known and highly regarded as a natural medicine. Aboriginal peoples have used the leaves in poultices and bandages for wounds, as an anti-inflammatory and antiseptic, and to aid decongestion. However, tea tree only began to become popular beyond its natural range around the 1920s. Later, during the Second World War, production of tea tree oil was considered essential war work in Australia to help treat wounded soldiers, and all Australian servicemen carried tea tree oil with them to help reduce the risk of infection. In more modern times it has had a great resurgence in popularity since the 1970s, and today many people around the world will have some kind of tea tree product in their homes, whether it be the pure essential oil, an antiseptic, or simply cosmetic face wipes or washes.

Although there are 256 species of *Melaleuca*, the essential oil known as tea tree oil is usually extracted from the narrow-leaved tea tree, *Melaleuca alternifolia*, which grows wild near streams and swamps in southern Queensland and northern New South Wales. It belongs to the myrtle family (Myrtaceae), which also includes eucalyptus and clove, from which of course other pungent oils are extracted. This species can grow up to about 7 metres (23 feet) high, but is most often seen as a tall, rather straggly shrub. In spring it produces compact clouds of tiny creamy-white flowers, which once pollinated develop into small woody fruits neatly arranged along the stems. The tea tree's bark is characteristically papery and peels away, leading to another of its common names in Australia – paperbark tree.

Tea tree is now grown commercially in large plantations. The oil is extracted by steam distillation from the narrow linear leaves, which are harvested in spring and summer. Studies into the effectiveness of tea tree derivatives are rare, but it is claimed that the terpinen-4-ol,

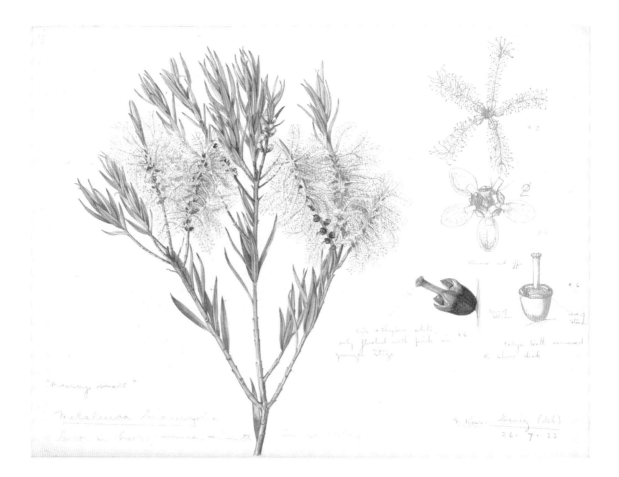

the oil produced by *M. alternifolia*, is a useful and safe anti-microbial product for use in both human and veterinary medicine. It is also said to be effective against certain fungal infections, and to be an anti-inflammatory, so it is not surprising that with such potential benefits so many tea tree products are now appearing on the market. And as useful alternatives to antibiotics continue to be sought, natural products such as tea tree oil will no doubt become more important in the fight against infectious diseases.

Ebony, *Diospyros ebenum*

BODY and SOUL

As well as providing food and spices, medicines and poisons, trees can also nourish our bodies and souls in a range of intriguing ways. They can represent life, but also death; they can embody spirits and ancestors, or be the basis of stories about creation and protection, gods and demons. Some species of tree, including the baobab in Africa and the Indian banyan fig in Southeast Asia, are revered by local people as being central to the life of the village. They offer shelter from the sun and weather, useful products and a place to meet to discuss important issues and take part in ceremonies. Many trees are linked to different religions, appear in traditional stories to entertain and educate, and have associated vivid symbolism. The hawthorn is perhaps a modest-looking tree, but is linked to numerous superstitions and folk tales, as well as to both pagan and Christian beliefs.

More practically, other trees are the source of resins and dyes used to decorate the body or create powerful and sensuous perfumes. A bright dye derived from the seeds of annatto is used as body paint and is also found in a wide range of foods and cosmetics today. When wounded, the dragon's blood trees yield a red resin that has both decorative and functional purposes. The sought-after frankincense tree has been at the centre of a worldwide trade in perfume and incense that have been essential in religious and other rituals for thousands of years. Trees can even help clothe us: white mulberry leaves are the favoured food of the insect that produces the raw material for that most luxurious of textiles, silk.

The wood of certain trees such as ebony was once vital in creating the musical instruments, from guitars to pianos, that lift our spirits and speak to our souls, and for pieces for playing games such as chess that engage our brains. Coffin tree wood is durable and resistant to decay, and as well as being favoured for building houses and temples was greatly valued in China for making coffins.

Natural tree products are far more embedded in our culture than we may realize. Like annatto, the soap bark tree is still used in traditional ways and is a highly valuable commercial crop important in the manufacture of cleaning agents and fizzy drinks. Few people read the ingredients on a lipstick, a bar of soap or a bottle of scent, and we rarely give much thought to how, where and when traditions have sprung up. However, the value of trees to many aspects of our lives and society should be recognized, as well as their role in making our experiences richer, more spiritual and more vibrant.

Baobab

Adansonia digitata

An iconic tree of the African landscape, the baobab is credited with being nourishing, medicinal and magical. Its immense girth and distinctive shape make it immediately recognizable – it is often known as the 'upside-down tree' as its bare branches in the dry season make it look as if it is growing with its roots in the air. It has attracted numerous other strange names too, including monkey bread tree, dead rat tree and the chemist tree. Baobabs are slow growing and long lived: one huge individual in Namibia (known as Grootboom) was radiocarbon dated as 1,275 years old. At the time of its collapse in 2004, this made it the oldest known flowering plant in the world, as most other ancient trees are coniferous, such as the redwoods, pines and yews.

Baobabs are wrapped in myth and legend, often featuring in creation stories, and are also perceived to have magical properties.

This species of baobab is native to the drier parts of tropical and southern Africa, where it is a widespread and common sight in the thorny woodlands of the savannah. A very majestic tree, it can grow up to 30 metres (100 feet) tall and the same in circumference, with an extensive root system. The disproportionately large trunk is an evolutionary marvel, allowing the tree to store water to survive through the many hot dry months which bake the arid and semi-arid landscapes where it grows. Large individuals are said to be able to store up to 100,000 litres (26,400 gallons) of water in their trunks, and elephants have been known to gouge them to reach the precious liquid inside. Many other animals also visit the baobab for food and shelter, including baboons and warthogs, as well as different types of birds, reptiles and insects.

The beautiful white petals of the baobab first unfurl in the evening and the flowers often only remain open for twenty-four hours. They are pollinated by bats and bush babies.

The baobab's large and showy white flowers appear at the onset of the rainy season, suspended on long pendulous stalks so that they can be more easily pollinated by bats. Each has a tuft of stamens projecting

below the flower to ensure pollen is brushed on to the pollinators,
which also include bush babies. The flowers open for just one day,
often first unfurling in the evening. After pollination, the resulting
velvet-skinned fruits are up to 35 centimetres (14 inches) long and are
filled with a dry pulp holding many small dark-brown seeds.

People have long valued the baobab. Some groups planted seedlings
near villages for the benefit of future generations, showing how integral these trees are to rural communities. The leaves, bark, fruit, roots
and seeds are all used in an extraordinary variety of ways. The tangy
pulp of the fruit has high levels of vitamin C – more than seven times
that of an orange – and is full of nutrients including calcium and potassium, as well as dietary fibre. Pulp and seeds can be made into drinks,
sauces, jams, soups and porridges, while the leaves are eaten as a staple
vegetable. The fibres from the inner bark are turned into rope, string,
mats, baskets, clothing and even beehives; dyes are extracted from the
roots, and the pollen from the flowers can be used as an ingredient
in glue. In all, more than 300 traditional uses of the African baobab
have been recorded. The hollow trunks have been converted into water
tanks, a prison, bars and tombs. Recently, oil from the seeds has been
evaluated as a biofuel and is also now used in moisturising cosmetics.

BODY AND SOUL

This useful tree also features in traditional medicine, with the fruit, seeds and leaves used in treatments for all manner of ailments including malaria, dysentery and diarrhoea, as well coughs and toothache. New research has looked into the anti-microbial, anti-viral and anti-inflammatory properties of the baobab, with many studies showing positive effects.

A baobab tree is often also a focus for village life – its shade is a good place to meet and resolve problems. Baobabs are wrapped in myth and legend, often featuring in creation stories and are also perceived to have magical properties. It is thought that spirits live inside the flowers, which should not be picked, while a drink made from the seeds is thought to protect you from crocodiles, making this a truly important and multipurpose tree.

The coffin tree

Taiwania cryptomerioides

At home in East Asia, this tree is among the largest species of the 'Old World forests' and a strong contender for the title of the world's tallest tree. In fact it closely resembles in stature, shape and form the current record holder, the Californian coast redwood (*Sequoia sempervirens*; p. 216). Some coffin trees are reported to be up to 90 metres (295 feet) tall, with soaring, straight trunks 4 metres (13 feet) or more in diameter.

It was first collected scientifically as a new species of conifer in 1904 by a Japanese botanist, Nariaki Konishi. Two years later, in 1906, it was described and published by another eminent Japanese botanist, Bunzo Hayata, who named many plants in the Taiwan region in the early twentieth century. As it was originally believed to grow only on the island, it was given the generic name *Taiwania*. However, it has since been discovered in parts of mainland China, Myanmar and northern

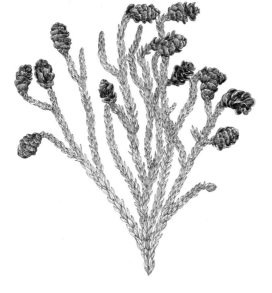

Vietnam. These mainland trees have been classified as a distinct species, *Taiwania flousiana*, but there is not thought to be any real botanical difference between the two species and the name is only used to distinguish the separate natural locations. *Taiwania* grows on several remote mountains on its home island at altitudes between 1,800 and 2,500 metres (5,900 and 8,200 feet), as well as on the western slopes of Mount Morrison, which at 3,952 metres (12,966 feet) is the highest mountain on the island and was named after the nineteenth-century English missionary Robert Morrison. Today it is known as Yushan, which translates as Jade Mountain.

Taiwania are truly magnificent denizens of this mountain, and they must have been an even more impressive presence before many were felled for

OPPOSITE AND RIGHT
Mature specimens of the
coffin tree have a tall, bare
trunk that soars upwards
to make this tree one of the
world's tallest. The timber
is highly resistant to decay
and was used for building
temples as well as for making
coffins. The tree's very
sharp bluish-green needles
are held in flat sprays.

There is now a ban on logging *Taiwania* in China, and the tree was celebrated in one of five Chinese stamps issued to highlight the protection that these trees deserve in their natural habitat.

their timber in the early 1900s. The most graceful and elegant of all the conifers, young *Taiwania* form a perfect pyramidal shape. The slender branches are drooping, with the growing tips curving upwards, and bear curtains of hanging branchlets in flat sprays of very sharp, glaucous, bluish-green needles that stand out clearly, especially in winter. It is also a long-lived, slow-growing tree species, which can reach up to 2,000 years of age. Mature specimens have a bare trunk towering above surrounding trees and a sometimes rather ragged canopy.

The timber of *Taiwania*, when seasoned, is light and soft and has a highly spicy but pleasant scent. It is extremely durable and resistant to fungal and insect attack and to decay, and in the past has been in great demand and overexploited for building temples and houses. Some trees have timber with a fine grain that is beautifully marked with red and pale yellow annual rings, making it very desirable for making high-quality furniture and other products. Perhaps most famously the wood has been used to make coffins, hence the common name. It was particularly sought after for the coffins of wealthy Chinese, and large trees could be extremely valuable. The English plant collector and explorer Frank Kingdon-Ward described in detail being taken to be shown a fine specimen of a coffin tree, and how trees would be cut down and sawn into planks before being transported. He estimated that a full-sized tree would yield sixty to eighty planks.

This stately tree was introduced to Western gardens by Ernest Wilson in 1918 and it is now grown as an ornamental for its attractive attributes. It is often assumed that as it originates in Taiwan it is not hardy enough to be planted outdoors without protection in northern latitudes, but it survives most winters and is much hardier than originally thought when grown in a sheltered, sunny position. In warmer regions including the Pacific Coast and southeast of the United States, Australia and New Zealand it grows much more successfully, though all the trees in these places are still relatively young.

In its natural habitat the coffin tree is recorded as vulnerable and it is threatened by overexploitation and illegal logging for its valuable timber. However, the establishment of Yushan National Park in Taiwan in 1984 and a logging ban in mainland China have given this remarkable tree some protection. In 1992 a set of special postage stamps was issued featuring five Taiwanese endemic conifers, including *Taiwania*, to highlight the protection that these rare and beautiful trees deserve.

Dragon's blood tree

Dracaena cinnabari and *Dracaena draco*

———

It has been prized since ancient times for its ruby red resin known as dragon's blood, which oozes from fissures and wounds in its bark.

A truly extraordinary and distinctive species, the dragon's blood tree *Dracaena cinnabari* grows only on the island of Socotra, 240 kilometres (150 miles) off the coast of the Horn of Africa. Often described as the most alien-looking place on Earth, this isolated island in the Indian Ocean is renowned for its otherworldly plants, including the desert rose (*Adenium obesum* subsp. *socotranum*) and the cucumber tree (*Dendrosicyos socotranus*), each having evolved strange but ingenious ways of surviving in the arid landscape. Socotra separated from mainland Africa millions of years ago, and around 37 per cent of its flora is endemic, that is it is found nowhere else. Of these species, the dragon's blood tree is perhaps the most famous.

This slow-growing evergreen tree has an umbrella-like form, or perhaps could be compared to an enormous green mushroom. The symmetrical and tightly packed domed canopy of long, sword-shaped waxy leaves helps conserve water in the hot, dry conditions. It also provides dense shade to reduce evaporation and channels any water to the tree's roots. The leaves are generally found only on the upper sides of the knotted branches, adding to the tree's eccentric appearance.

Once common across Socotra, healthy populations of these trees are now mainly to be found at higher altitudes in and around the Haggier Mountains, particularly on the limestone plateau of Rokeb di Firmihin. The dragon's blood tree requires certain amounts of rain, mist and cloud cover in order to thrive in the harsh, rocky soils, and these conditions have become rarer over the past century. Seedlings often fail to grow where there is no understorey of shrubs to protect them against drought, and grazing by goats and cattle can reduce the

Mr Smith.
May 31 - 1913

copy of a drawing sent by
Dr George V. Perez
Puerto Orotava Teneriffe

Dracaena Draco

habitat of this emblematic species. Climate change is also playing its part in its decline, and it is now classed as 'Vulnerable' on the IUCN *Red List of Threatened Plants*.

Clusters of greenish-white, fragrant flowers appear at the ends of the branches in February, followed by small fleshy fruits that turn an orange-red when they ripen. These attract birds and animals, which eat them and disperse the seeds; the loss of any of these species affects the ability of the tree to regenerate and spread. It is not only these other creatures that value this tree, as it has also been prized since ancient times by people for its ruby red resin known as dragon's blood. The red sap oozes from fissures or wounds in the bark to help protect the tree from infection, and has been harvested and put to many uses over the centuries – including as dyes and paints to decorate objects and houses as well as for body adornment, and has also had a role in local traditional medicine. Once widely and regularly harvested, the resin is in less demand today.

Dracaena cinnabari is closely related to *D. draco*, the dragon tree of the Canary Islands, Cape Verde and Madeira, which it closely resembles. It too was once extensively used for its resin, while its fruit is said to have been the food of a now extinct dodo-like bird. Unfortunately this dragon tree has also experienced a severe decline in numbers in the wild as a result of habitat loss and grazing.

Ebony

Diospyros ebenum

ABOVE The black heartwood of ebony is extremely hard, allowing intricate decoration to be carved on objects made from it, such as this comb.

OPPOSITE Different types of ebony grow in various regions of the world, but *Diospyros ebenum*, the Ceylon or Indian ebony, is one of the most highly regarded for its wood.

Ebony is a tree that has given us beautiful music and has graced the finest furniture. Craftsmen have sought out its deep black, fine-grained wood for hundreds of years, and it has been incorporated into pianos, violins and cellos, and used as inlay and veneer for the very best cabinets, tables, chairs, clocks, as well as chess sets and many other decorative objects.

The timber of ebony is extremely durable, being both heavy and very dense – so much so that some types do not float in water. Such hardness makes it difficult to work, but means an exquisite glossy finish can be achieved, comparable to marble, and it allows extremely fine detail to be carved into it. Considered to be the best of all cabinet-making woods, ebony has been revered for centuries in Europe. In the fifteenth century, carpenters in Germany specialized in making fine cabinets from it, and in France in the sixteenth century cabinet makers were defined by their work in ebony and were called 'menuisier en ébene' (workers of ebony) or 'ébenistes'. The addition of ebony to any piece marked it out as a luxury item, available only to royalty and the extremely wealthy.

The qualities of this black, hard wood were certainly known about and admired in Britain in Shakespeare's day, as he includes references to it in his plays, including *Love's Labour's Lost*, believed to have been written in the mid-1590s. Ferdinand, King of Navarre, exclaims: 'By Heaven, thy love is as black as ebony', to which the nobleman Berowne replies: 'Is ebony like her? O wood divine! A wife of such wood were felicity.'

Several species of *Diospyros* were exploited commercially as 'ebony', and sometimes other genera too, but Indian or Ceylon ebony, *D. ebenum*, and the Coromandel ebony or tendu, *D. melanoxylon*, were regarded as the best and most valuable trees. Indian ebony was thought to have the most reliably black heartwood – others could be streaked with brown

J. Prevsne del.   N^{elle} Noiret & Mougeot sc.

Plaqueminier Faux-Ebénier.

and grey. This evergreen, slow-growing species, which can reach up to 20 metres (66 feet) tall, is native to the humid coastal forests of Sri Lanka and southern India. Other ebonies grow widely across Southeast Asia, southern India and some parts of tropical Africa.

Ebony timber was always a rare and expensive item. It was imported into Britain in the 1600s from India, Mauritius and Madagascar, but it was not until the 1800s that access to African markets and a British presence in Sri Lanka meant that supplies increased and the wood became more widely used. By the 1900s, however, sources of the best ebony were becoming scarce, and early in the twentieth century shipments from the East ceased. Even though it is such a highly regarded tree, it was not managed in a sustainable way and is now regarded as under threat of extinction. Today, trade in the timber of many ebony species is restricted, although illegal trade continues to be worth billions of pounds.

'Is ebony like her?
O wood divine! A wife of
such wood were felicity.'

Diospyros of different kinds are valued not only for their timber. The Indian ebony, for instance, produces small golf-ball-sized fruits which are a rusty velvet on the outside and deliciously edible. Some surprising relatives are grown specifically for their fruits, including persimmons, notably *D. kaki* and the American *D. virginiana* (p. 99).

While ebony timber is highly sought after and valuable, species of *Diospyros* also produce edible fruit. Those of the Indian ebony are golf-ball-sized and reputed to be delicious.

Indian banyan fig

Ficus benghalensis

To encounter an old Indian banyan fig is to meet a giant of the botanical world as well as a symbol of faith and society. Some of the largest specimens can have a spread of up to 200 metres (656 feet) in diameter and be 30 metres (100 feet) tall. Their wide, sprawling trunks, complete with astonishing dangling aerial roots, evoke a vision of the mysteries of India, and indeed this species is India's national tree.

Visitors to these trees often leave ribbons or small figurines tucked away among the roots and branches as symbols of their hopes and prayers.

Vast banyan figs, with their large leathery leaves, offer welcome shade in their native India and Pakistan, and they are often used as the setting for market trading and as meeting places. Many grow naturally in the wild, but they are also planted in parks and along streets in India and near Hindu temples. Visitors to these trees often leave ribbons or small figurines tucked away among the roots and branches as symbols of their hopes and prayers – for this reason the banyan fig is known as the 'wish-fulfilling tree'. It is sacred in the Hindu religion, and the roots, trunk and leaves are associated with the deities of creation (Brahma), stability (Vishnu) and destruction (Shiva). The banyan fig also represents eternal life, which perhaps is not surprising given the way in which this species continues to grow and spread.

What at first sight might appear to be a forest of different trunks, with vast tangled curtains of thin aerial roots, is often in fact one single ancient tree. This individual would have begun life as a seed from one of the banyan's bright red figs that had been deposited by an animal or bird in a crevice in the branches of another tree. Once large enough, the seedling sends down roots towards the ground, eventually creating a network running down the trunk of the host tree. Over time the fig's roots fuse together and completely encase the tree, ultimately killing it by depriving it of light and nutrients, thus becoming its shroud. This growth habit makes the banyan one of the 'strangler' figs.

The banyan fig continues to send down aerial roots from its branches throughout its life, which in turn become like trunks and support the tree as it stretches out ever wider. A survey of the 'Landmark Trees of India' found seven enormous banyan figs, with the largest of all, called Thimmamma Marrimanu, growing in the village of Anantapur around 160 kilometres (100 miles) north of Bangalore. This magnificent tree spreads over 2 hectares (almost 5 acres), with 4,000 prop roots, forming one of the largest tree canopies in the world. A small temple is located near its centre, and both tree and temple are visited by people wanting to be blessed, especially childless couples.

However, this species is not just impressive and sacred – it is also extremely useful. The sap is an ingredient in many traditional medicines for treating skin complaints, from cuts to blisters and bruises, as well as being used to ease toothache, as an aphrodisiac, for promoting hair growth and for snake bites. It has also been incorporated in remedies for coughs, vomiting, diarrhoea, and even for more serious

OPPOSITE The banyan fig is a strangler fig: it germinates in the branches of another tree and sends down aerial roots which slowly encase the host. Many large old specimens are used as meeting places and are valued for the shade they provide.

BELOW Indian banyan fig trees are sacred in the Hindu religion and temples are often built near or even in them.

BODY AND SOUL

diseases such as cholera and dysentery. Today, it is being investigated for its potential in the treatment of diabetes.

Banyan timber is hard and can be utilized for all manner of everyday products, while rope and paper are made from fibres from the bark. Perhaps one of the most surprising products arising from this giant tree species is 'shellac', an important ingredient of French polish. This valuable harvest is derived from a resinous secretion produced by female lac insects living in the tree, and has a wide variety of uses. As well as polish, shellac is found in varnishes, primers and wood-finishing products, and also in medicines, sweets, clothes, cosmetics, wristwatches and even fireworks, and old gramophone records were once made from it.

Frankincense

Boswellia sacra

One of the first and most famous Christmas presents ever given is surely frankincense. This aromatic resin has scented the air and perfumed our lives for millennia. At the time it was presented as one of three gifts at the birth of Jesus, as recorded in the Bible, the resin of the frankincense tree had a value equal to or greater than another of those gifts, gold. Frankincense has in fact been traded for up to 5,000 years, probably mainly originating in the Arabian Peninsula. It was an essential component of the rituals of many religions, and the ancient Egyptians imported it for use in their temples and mummification processes and as a medicine.

Several species of *Boswellia* can yield the precious resin, but the most highly regarded, *B. sacra*, is native to southwest Oman and southern Yemen, as well as Somalia and Ethiopia in northeast Africa. A small deciduous tree, up to 8 metres (26 feet) tall, it grows in dry scrubby woodlands on limestone slopes, as well as coastal hills which are regularly clouded in fog during the summer. Usually multi-stemmed

and with a broad, cushion-like base thought to help with stability in the harsh rocky terrain, the trees have papery, peeling bark and tangled branches. If the bark is wounded, an oily gum-resin is exuded by the tree as a barrier to prevent anything noxious entering. The resin runs down the trunk in milky-white droplets or 'tears', which then harden to a red or golden-brown colour.

The area where wild frankincense trees still grow today includes a UNESCO World Heritage Site in Oman known as the Land of Frankincense. This region was once the beginning of the Incense Trail – an overland route thronged with traders transporting

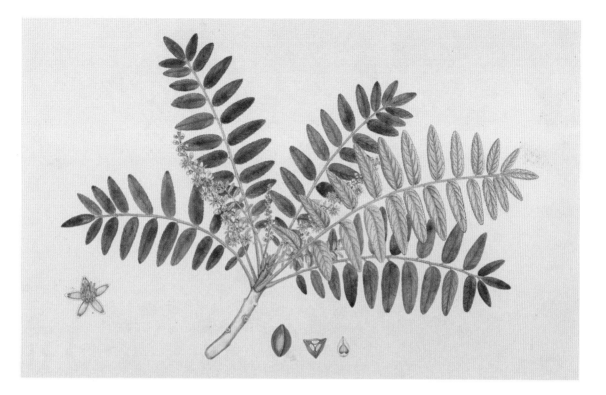

the resin to the markets of Mesopotamia and the Mediterranean, and linked with other major routes for the movement of goods such as the Silk Road. Frankincense brought huge wealth to Arabia and beyond – the ancient city of Petra in Jordan was built by the Nabataeans with wealth created by the incense trade. It was of such importance that several classical authors including Herodotus, Theophrastus and Pliny the Elder mention its origins, harvesting and uses.

Frankincense is still harvested in the traditional way from cuts made in the bark. The resin begins slowly to ooze from the wound, taking several days to form into large droplets which can be cut off and are dried for a couple of weeks; it is then ready for sale. Despite the small size of these fragrant nuggets, a single tree can yield several kilograms a year. After centuries of sustainable harvesting, a recent rise in global demand from the perfume industry now threatens to damage wild trees in places such as Somalia.

As well as being used for incense, frankincense has been dissolved in water to treat fevers and coughs, ulcers, nausea and indigestion among other ailments. Today it is still used in deodorants and toothpaste in Oman, as well being burned to drive away mosquitoes. More romantically, legend has it that the phoenix builds its nests in the branches of the frankincense tree and feeds upon its tears.

Hawthorn

Crataegus monogyna

A small, shrubby and very thorny tree, often seen in the hedgerows of northern Europe, the hawthorn can easily be overlooked as an ubiquitous part of the countryside. It may be overlooked, but the hawthorn has long been entwined with history, cultures, folklore and religions, dating back into prehistory. It is also a vital part of the natural ecology of the landscape, dotting woods and hedges with a welcome profusion of blossom in spring and an abundance of red berries in autumn.

> *The hawthorn has long been entwined with history, culture, folklore and religions, dating back into prehistory.*

The hawthorn is also known as the mayflower as May is the month in which it produces its billowing clouds of small white flowers, tinged with rose-pink, which are held above the fresh green, lobed leaves. It is closely associated with May Day (1 May) – the ancient Celtic festival of Beltane and today a holiday in many countries. In pagan pre-Christian times the May Day festival was a raucous affair, but by the nineteenth century it had become far more sedate and romanticized, with dancing round the may pole and a Lord and Lady of May being crowned each year by upstanding members of the community. From 1881, the English art critic John Ruskin presided over May Day celebrations at Whitelands College in Chelsea, and he would present the beautiful Whitelands Cross, a gold hawthorn sprig, to an elected May Queen.

Hawthorns are invaluable for wildlife – offering nectar in spring for insect pollinators, leaves for caterpillars and fruits or 'haws' in autumn for many birds and mammals, as well as providing a protective environment for nesting birds. This illustration shows shrikes in a hawthorn from John James Audubon's *Birds of America*.

For a modest-sized tree, growing up to 15 metres (50 feet) high, the hawthorn has attracted an impressive number of superstitions. For instance it was thought that fairies used these trees as their homes, and in Ireland and elsewhere it was considered bad luck to harm a lone hawthorn, or to bring its flowers into the house as this could be an omen of illness or even death. The flowers' bad repute may be linked

PLATE. CXCII.

Great American Shrike or Butcher Bird. LANIUS SEPTENTRIONALIS. *Male, 1.2. Summer Plumage. Nº 3, Young or Winter Nºs 4. & Crataegus apiifolia.*

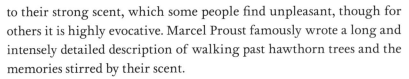

to their strong scent, which some people find unpleasant, though for others it is highly evocative. Marcel Proust famously wrote a long and intensely detailed description of walking past hawthorn trees and the memories stirred by their scent.

Hawthorns are also wishing trees (or 'clootie' trees): passers-by tie a ribbon or piece of cloth around the branches, or push a coin into the bark, as a wish for good health, love or help in their lives. Such trees, often next to holy wells, soon became renowned, and many were to be found on paths trodden by pilgrims on their way to seek redemption at cathedrals or other sacred sites. At Beauraing in Belgium is a shrine to the Virgin Mary where she was said to have appeared to children beneath the branches of a hawthorn tree.

The most famous hawthorn in Britain is the Glastonbury Thorn in Somerset. A cultivar of *Crataegus monogyna* called 'Biflora', it flowers twice a year, as its name suggests, including once around the time

of Christmas. According to the story told by the monks of Glastonbury Abbey, the original hawthorn sprang from the staff of Joseph of Arimathea, who, having witnessed the crucifixion of Jesus, travelled to Britain as a missionary and pushed his staff into the ground when he began to preach. It immediately took root and flowered. Many trees have since borne the name of the Glastonbury Thorn and recent plantings are said to be grafts from much older trees. One of these was cut down by Oliver Cromwell's men during the English Civil War in the seventeenth century because it was seen as a symbol of superstition, and another was felled more recently, in 2010, by unknown assailant with a chainsaw.

The hawthorn is a member of the rose family (Rosaceae) and there are estimated to be more than 200 species of hawthorn found in the northern temperate zone, from China to America. The longest thorns belong to the American cockspur thorn, *Crataegus crus-galli*, which are over 7 centimetres (3 inches) long. *Crataegus monogyna*, the common hawthorn, is integral to the history of the countryside of northern Europe. It has traditionally been used as a hedging tree, as it makes a fast-growing (another English name for it is quickthorn) and thorny barrier to keep farm animals in their fields and anything else out. It is also invaluable for

wildlife, offering nectar in spring for insect pollinators, leaves as food for caterpillars and its fruits or 'haws' in autumn for birds and small mammals such as the dormouse, as well as providing protection for nesting birds.

Although hedgerows are such an important part of the fabric of the rural landscape they did not fare well in the last century, with thousands of miles ripped out after the Second World War to create bigger fields that could be more easily worked by large machinery. And so the hawthorn is not as common as it once was. Thankfully, many people began to realize the damage that had been done to both landscape and wildlife, and this iconic small tree is now a symbol of a well-managed countryside.

Mulberry

Morus nigra and *Morus alba*

———

There are about twelve species of mulberries, all of which are deciduous flowering trees in the family Moraceae. This large family also includes tropical trees such as the breadfruit (*Artocarpus altilis*) as well as the edible fig (*Ficus carica*) and its numerous relatives. Many of the mulberries grow in tropical regions of Africa and the temperate parts of Asia and North America, but two important species commonly grown in gardens today are the black mulberry, *Morus nigra*, which originates in Central Asia, and the white mulberry, *Morus alba*, from China, where it has been cultivated as an essential part of the silk industry (sericulture) for over four thousand years.

The male catkins ... are noted for their rapid release of pollen. The stamens act like a catapult and fling pollen at ... over half the speed of sound.

The white mulberry is a fast-growing, small to medium-sized tree with a sprawling crown and a rugged trunk. Its leaves can vary markedly in size and shape, from smooth to deeply lobed on one side only or both. The male and female flowers are usually found on separate trees, and the male catkins have one surprising accomplishment – they are noted for their rapid release of pollen. The stamens act like a catapult and fling pollen at approximately 560 kilometres (350 miles) per hour – over half the speed of sound, making it the fastest known movement yet observed in the plant kingdom. The fruits are like blackberries in shape and change in colour from white when young to burgundy-red or purple when ripe. They are said to be toxic when immature and when ripe are bland in taste compared to the black mulberry.

The black mulberry is a long-lived, stately tree. With age, the gnarled trunk often leans at an angle and has to be propped up to provide support. The spreading branches with heart-shaped leaves

The black mulberry originates in Central Asia and is now often grown in gardens for its delicious fruit. Old trees can have very gnarled bark and branches that droop to the ground, where they may take root.

BODY AND SOUL

Plate 126.

The Mulberry Tree
Eliz. Blackwell delin. sculp. et Pinx.

1. Cluster of Flowers
2. Flower separate
3. Fruit
4. Seed

Morus - nigra vulgaris

sometimes droop to the ground and bear the succulent dark purple, almost black, raspberry-like fruits with a unique, sub-acid taste that this tree is often grown for.

These two mulberries are frequently confused in cultivation and sometimes are wrongly planted for their intended use. This is exactly what King James I of England did in the seventeenth century when he wanted to compete with the silk industries of Italy and France. Thousands of mulberry trees were imported and the king had his own 1.6-hectare (4-acre) orchard planted in a garden north of present-day Buckingham Palace. The trees were cultivated by what were known as the 'King's Mulberry Men'. In 1609 James wrote to his lord lieutenants to encourage them to plant groves of mulberries as a food source for the silk worm (*Bombyx mori*) which produce the silk threads for weaving into this delicate cloth. Unfortunately, it was the black mulberry that

OPPOSITE The white and the black mulberry, both shown here, are often confused. The fruits of the black mulberry are a real delicacy, while the leaves of the white mulberry are the staple diet of the silk worm.

BELOW The white mulberry has been cultivated in China for thousands of years for silk production. Having fed on the leaves of the tree, the silk worm forms a cocoon; these are collected and the fibres are spun and made into silk. The process now takes place in industrial facilities.

was planted: although the silk worms will feed on the black mulberry, they much prefer the leaves of the white mulberry (something the French were aware of) and the silk industry failed in Britain.

It may have been a mistake, but if so it was a fortunate one, as the black mulberry grows better in Britain and had in fact been cultivated there well before King James's day. The Romans considered the fruits a real delicacy and also used them in medicines. Today they are still highly regarded for conserves and drinks, and are popular with ice-cream makers and gin distillers. A word of warning, however – they are impossible to pick without the juice oozing from the ripe fruits, which will permanently stain any garments blood red.

The Roman poet Ovid explained the origin of the fruit's striking colour in the fourth book of his *Metamorphoses*, which William Shakespeare then incorporated in *A Midsummer Night's Dream*. In this tale of forbidden love, Pyramus and Thisbe plan secretly to marry beneath a mulberry tree. Thisbe, arriving first, flees from the meeting place after a lion appears; she drops her scarf, which is torn and bloodied by the beast. Pyramus then discovers the bloodstained scarf, and, supposing his beloved to be dead, he stabs himself and his blood stains the white fruits of the tree a dark red. From that day on, the juice of the fruit remained a deep, ruby red.

Soap bark tree

Quillaja saponaria

In the dry forests of central Chile grows a tree that has in the past provided a very useful product to the peoples of the Andes and is still today carefully harvested as a highly valuable commercial crop. As indicated by its Latin specific name, *saponaria*, and its English common name, the inner bark of the soap bark tree, when dried and ground, can be used to make a natural gentle soap. It contains a saponin, which foams when mixed with water and can be worked into a lather to clean all manner of things.

As the saponins derived from this tree are safe, consistent and effective, they are now found in a wide variety of products, including soaps and shampoos. Perhaps more surprisingly, extracts from the soap bark tree are also used as a foaming agent in fizzy drinks and root beers, and an ingredient in foods including desserts and sweets. They are even used in fire extinguishers and agricultural crop sprays and were previously a component in the developing fluid for photographic film. The many uses of the soap bark tree do not end there, however, as the saponins have recently been shown to increase the efficacy of certain animal vaccines. A highly purified extract was tested as an 'adjuvant' in these vaccines and results showed that adding just a tiny amount increased the power of the vaccine.

The soap bark tree is a medium-sized evergreen tree with attractive white flowers. The genus, *Quillaja*, is named after the local Chilean word for it. In Chile, where it is cultivated and harvested, growers are careful never to take more than 35 per cent of a tree on a five-year cycle. Such pruning encourages new growth and creates a sustainable crop. In its native land the tree has a long history of use in traditional medicine as a treatment for chest problems. Modern medicinal uses of soap bark continue to be investigated, so new products or medicines may yet come from this natural Chilean treasure.

The Chilean soap bark tree has large, leathery evergreen leaves and delicate white flowers. Its bark contains a saponin, which can be extracted and made into a natural soap but also finds use in cleaning products, fizzy drinks and even vaccines.

Annatto

Bixa orellana

Strange as it may seem, a South American tree with bright orange-red seeds can link certain cheeses and lipstick, and indeed many other familiar foodstuffs and cosmetics. Annatto or achiote is a small evergreen tree which can grow up to 30 metres (100 feet) high. It has wide, heart-shaped leaves and whitish-pink flowers with a prominent boss of pink stamens. Once pollinated, the flower develops into a spiky fruit capsule that splits open when ripe to reveal dozens of angular red seeds. It is these seeds that yield a useful dye, as they contain a soluble pigment known as bixin. Today, bixin is second in importance only to saffron (economically) as a natural colourant.

> *The ancient peoples of South America used annatto for hundreds of years as a body paint as well as to colour and spice up food. The Aztec added it to their chocolate drink.*

The ancient peoples of South America used annatto for hundreds of years as a body paint as well to colour and spice up food. The Aztec added it to their chocolate drink, while also in Mexico annatto was used as a source of red ink for manuscript painting in the sixteenth century. Such traditions continue, as indigenous peoples in South American countries still use the vibrantly coloured seeds to dye hair, clothes and food, or mix them with oil to apply as a body paint. As well as being decorative, a covering of annatto can also act as a sunscreen and an insect repellent. It has been an ingredient in traditional medicine for centuries too – the bark, leaves and seeds have all been used in treatments for a variety of common illnesses and conditions, for instance sore eyes, bruises and wounds, as an expectorant and even as a laxative. The tree is so useful that it has been given many different names across the regions in which it now grows.

Annatto has been traded since the seventeenth century and in the eighteenth century it was used in Spain as a dye in the silk industry.

Its specific name commemorates Francisco de Orellana, a Spanish explorer and conquistador. Today annatto is widely cultivated in tropical areas – from its native Brazil to the Caribbean and even India and Sri Lanka. And around the world, many people will be consuming or using annatto without realizing it, as it is found in foods and cosmetics from cheese, butter and margarine, popcorn, custard, snack foods, sweets and cereals, to shampoos, skin care products and cosmetics – another common name for this tree is in fact the lipstick tree. As a crop, the natural red dye from this small forest tree has a vast global market and has made a huge commercial impact far beyond its native range.

ORLEANA, SEV ORELLANA FOLLICVLIS LAPPA   CEIS   Fig.

The spiky fruit capsules hold dozens of bright red angular seeds, which are the source of a widely used natural dye.

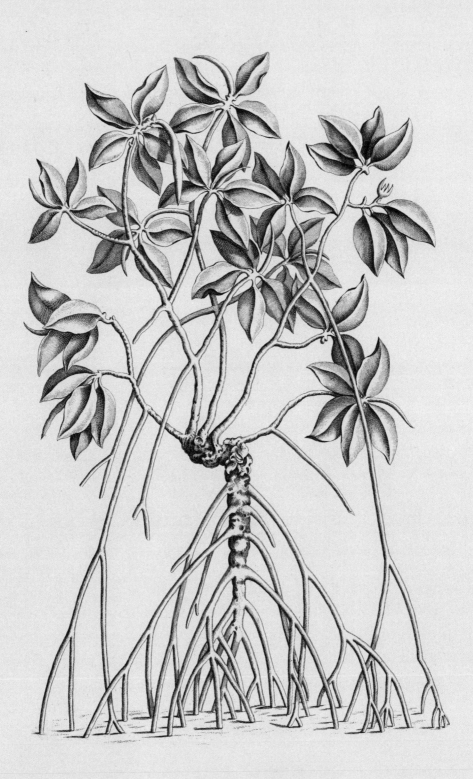

Mangrove, *Rhizophora mangle*

WONDERS OF THE WORLD

Trees exert a powerful influence over our imaginations. We are fascinated by the superlative trees and are amazed that anything could be so tall, old, massive, rare, delicious or beautiful. These are the trees that command our respect and admiration, and act as ambassadors for their kind. Some of the largest and most impressive trees in the world – such as the redwoods, kauri, alerce and eucalyptus – act as icons or national symbols of their native lands. They represent the history and diversity of the nature and the culture of those regions and offer us an insight into other worlds and times.

The coast redwood, Douglas fir and mountain ash have long vied for the accolade of being the tallest tree in the world, and individual superlative specimens earn names of their own. Currently the prize goes to a coast redwood called Hyperion, which stands at around 115.9 metres (380 feet) tall in Redwood National Park in California. But trees change all the time, and the thought that others nearby, still waiting to be measured, may be even taller is tantalizing.

Height is not everything of course. There are also records for the largest by volume – a giant redwood – and the heaviest, Pando, a quaking aspen, not to the mention the oldest, a bristlecone pine appropriately named Methuselah, although dating trees remains a source of contention among experts. That such individuals have lived through so many generations of our history, and grown taller and stronger while our life cycles are so relatively short seems to resonate deeply with us.

Some species have other curious or even bizarre attributes that captivate us, such as the durian, universally agreed to have the world's most foul-smelling fruit, which many people also find delicious. The extraordinary coco de mer, or sea coconut, holds several records, most notably for the largest and heaviest wild fruit. The mangrove astonishes us by being so highly adapted to the specific niche it chooses to grow in, tidal saltwater, where it is hard to believe anything could thrive. Others capture the spirit of adventure and exploration, from the distinctive banksias encountered by the first Europeans to set foot in Australia, to the sublime handkerchief tree collected in the wilds of China. The recent rediscovery of species thought long extinct by science, including the dawn redwood, has also caused worldwide excitement.

Fortunately, it seems interest in trees is growing and flourishing, and with it an awareness of how important they are to us. Long may trees continue to enchant and delight us.

Mountain ash

Eucalyptus regnans

The common English name for this tree, mountain ash, is rather misleading – it is not an ash at all. Other names, including swamp gum and stringy gum, are much more appropriate, as this is a species of *Eucalyptus*, a large genus of evergreen flowering trees and shrubs in the myrtle family (Myrtaceae), which are often known generally as gum trees. The eucalypts are a huge and diverse group, with over 700 species, and eucalypt forest makes up more than 92 million hectares (227 million acres) of the native tree flora of Australia. The different species are some of the most difficult trees to tell apart, as they are all very similar in appearance except to the well-trained taxonomist. Two other closely related genera once shared the name *Eucalyptus*, but in 1995 they were controversially split into their own separate genera. *Corymbia* is a group of around 113 species of bloodwoods, ghost gums and spotted gums, and *Angophora* contains about 22 species known as the rusty gums. But they are all still eucalypts.

The genus was first described in 1788 by Charles Louis L'Héritier de Brutelle, a French botanist and magistrate, who named the first eucalypt, *Eucalyptus obliqua,* the messmate stringybark. This specimen had been collected on Bruny Island, off Tasmania, by gardener-botanist David Nelson on James Cook's third expedition in 1777. It was brought back to the Royal Botanic Gardens, Kew, where L'Héritier was working at the time. L'Héritier's generic name is derived from the amalgamation of two Greek words, *eu*, meaning 'well', and *calyptos*, meaning 'covered', which describes the *operculum* or bud cap that conceals the flower before opening.

Eucalyptus regnans, the mountain ash, is native to Tasmania and Victoria in southeastern Australia, where it grows to heights of between 70 and 114 metres (230 to 375 feet), making it the tallest broadleaved tree in the world. Its tall, straight trunk is covered with a smooth grey

Native to Tasmania and Victoria in southeastern Australia, the mountain ash, a eucalyptus, is second only to the coast redwood as the tallest tree on Earth, but it is the tallest broadleaved tree in the world.

bark, except on the lower trunk and buttresses of the tree, where it is rough and fibrous. The glossy grey-green leaves change shape as the tree ages, becoming more lance-shaped as it becomes adult, when white flowers are also produced in late spring.

The mountain ash was described as the 'loftiest tree ... and stupendously tall' by Victorian botanist Baron Sir Ferdinand Jacob Heinrich von Mueller in 1871. He named the species *regnans* meaning 'ruling' in Latin, alluding to the height and dominance of these trees. The Ferguson Tree, found in the Watts River region of Victoria in 1871, measured 132.6 metres (435 feet) and is claimed to be the tallest recorded mountain ash, but the measurement was taken on the ground with a surveyor's tape after the tree had fallen and so is not regarded as reliable. The tallest known living specimen today is called Centurion. It was discovered in 2008, growing in a southern Tasmanian forest and measured as 99.82 metres (327 feet), making it the second tallest tree in the world after the coast redwood, *Sequoia sempervirens* (p. 216).

It can only regenerate from seeds, which are released from their woody seed capsules or 'gumnuts' by the heat of fierce fires, producing up to 2.5 million seedlings per hectare.

This species of eucalypt grows in pure stands in wet rainforest, and unlike many other eucalypts does not have a lignotuber, a woody swelling of the root crown containing buds from which new sprouts can grow after fires. As it cannot regrow in this way, it can only regenerate from seeds, which are released from their woody seed capsules or 'gumnuts' by the heat of fierce fires, producing up to 2.5 million seedlings per hectare, which are then naturally fertilized by the fire ash.

The common name mountain ash came about because the tree's wood is very similar in appearance to that of the true ash tree, *Fraxinus*. The long, straight, knot-free trunks yield a timber that is yellow to light brown in colour, with a good grain and very durable. Early settlers gave this tree another of its common names, the 'Tasmanian oak', because they compared the strength of the timber to that of the English oak, *Quercus robur* (p. 36). It is highly valued by architects, builders and furniture makers, and is logged for use as panelling, flooring, veneer and plywood and in general construction.

If grown as an ornamental tree outside its native range, *Eucalyptus regnans* is not totally frost hardy and needs a mild, temperate climate to survive and reach its full potential. Eucalypts have a bad reputation among horticulturists and garden designers as being fast-growing, thirsty trees with an unstable root system, which makes them susceptible to toppling in wind as they reach maturity. They are also very difficult to blend into a northern hemisphere tree-scape, with their

unique bluish foliage and unusual, attractive barks. The most commonly planted eucalypts since the mid-nineteenth century are the cider gum, *E. gunnii*, and the snow gum, E. *pauciflora* subsp. *niphophila*, but with climate change and predictions that temperatures will rise, there may well be a place in gardens for more species of these beautiful ornamental trees.

RIGHT The genus *Eucalyptus* was first described in 1788 by Charles Louis L'Heritier de Brutelle, and today includes over 700 species. They are all very similar in appearance, with mature trees usually having long, lance-shaped waxy leaves, which are evergreen.

OPPOSITE Eucalyptus oil is extracted from the leaves of gum trees by distilling with steam, and can be used as an industrial solvent, an antiseptic, a deodorant and, in small quantities, as an additive in sweets, cough drops, toothpaste and decongestants.

Alerce

Fitzroya cupressoides

A rare, slow-growing but long-lived conifer in the Valdivian rainforest of Chile and Argentina, and the largest tree species in South America, the magnificent *Fitzroya cupressoides* can reach a height of 60 metres (198 feet) with a trunk diameter of up to 5 metres (over 16 feet) across. In nature it makes a stately, single-trunked specimen, with a dense, unbalanced pyramidal canopy. It is usually found growing in moist forests at elevations of between 1,000 and 1,500 metres (3,280–4,920 feet), mixed with the southern beech, *Nothofagus*, and another rare conifer, the Guaitecas cypress (*Pilgerodendron uviferum*), for which it is often mistaken.

> *The wood of alerce was once even used as a form of money called Real de Alerce.*

The only species in the genus, it was named *Fitzroya* by Charles Darwin in honour of Robert FitzRoy, who was the captain of HMS *Beagle* on Darwin's five-year voyage to South America, including the Galapagos and Tierra del Fuego. It is believed that on this voyage Darwin came across a specimen of 12.6 metres (41 feet) in diameter. A common name for this impressive tree is the Patagonian cypress, but it is known more popularly and locally as alerce, which is Spanish for larch. As often happens with common names, this is confusing since the larch is a deciduous conifer while alerce is very much evergreen, having needle-like leaves held in whorls. Many of the national parks where it can be seen growing today incorporate the common name, for example Alerce Costero National Park in the Los Ríos Region near Valdivia and Alerce Andino National Park in the Los Lagos region of Chile.

The popular common name for this tree is confusing as 'alerce' is Spanish for larch, which is deciduous, while *Fitzroya* is a true evergreen, with short, needle-like leaves that have a graceful, pendulous habit on the tree.

Alerce has long been valued for its timber, which has been used in construction as well as for ships' masts and furniture, but after the arrival of the Spaniards in the sixteenth century the felling of trees greatly increased and it is now endangered. The islands of Chiloé were

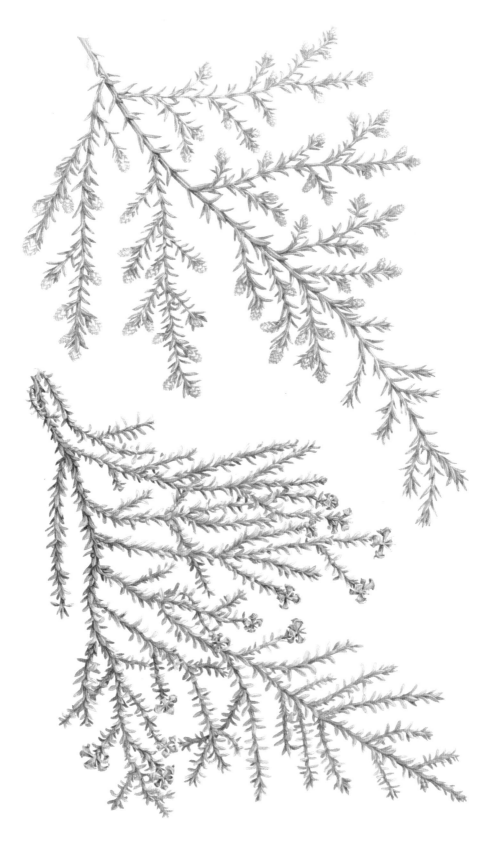

ALERCE

Unsustainable logging has resulted in this magnificent conifer becoming rare in its native habitat of the Valdivian rainforest of Chile and Argentina. It is the largest tree species in South America, reaching a height of 60 metres (198 feet).

once covered in dense forest where *Fitzroya cupressoides* was common, but today the trees are dead or scarce as a result of deliberately created wild fires and artificial drainage which has resulted in drought conditions. Many trees were also felled and converted into wood shingles for traditional buildings of Chilota architecture in colonial Chiloé. These very desirable shingles are resistant to rot and insect attack and were also the principal means of exchange in trade with Peru. The wood of alerce was once even used as a form of money called Real de Alerce.

Fitzroya are now rare in their natural habitat because of such heavy and unsustainable logging, particularly in the nineteenth and twentieth centuries. But there is some hope, as much conservation work is being done today in national parks in Chile and internationally in botanic gardens and collections to preserve this beautiful tree and prevent its threatened extinction. In 1973 alerce was protected under the Convention on International Trade in Endangered Species of Wild Fauna and Flora (CITES), banning the international trade in its wood. In 1976 the tree was declared a national monument in Chile and it became illegal to cut down living trees, though logging still occurs occasionally.

The largest known living specimen, Gran Abuelo or Alerce Milenario, was discovered growing in Alerce Costero National Park in 1993 and is thought to be nearly 3,644 years old. It is possible that some living trees are even older, but their rings cannot be counted as the trunks are hollow. This makes alerce the second oldest documented living tree species, after the bristlecone pine (*Pinus longaeva*) of the White Mountains of California.

Fitzroya was introduced into cultivation in 1849 by William Lobb, a Cornish plant collector for the Veitch Nurseries in Exeter, southwest England, and is now grown in Europe as an ornamental. If it can be given the perfect growing conditions of fertile, moist but well-drained soil in a sheltered position, it makes a small, multi-stemmed shrubby tree with fine bluish-green branches of drooping habit. But even though attractive in a garden setting, such specimens cannot compare to the size and stature of these magnificent trees in their natural environment in Chile.

Bristlecone pine

Pinus longaeva

The long cylindrical cones of the bristlecone pine take around sixteen months to ripen to an orange-buff colour. Each scale has a bristle-like spine and opens to release the seeds, which are dispersed mostly by wind, but also by a bird called the Clark's nutcracker.

Methuselah, a descendant of Seth, the son of Enoch, father of Lamech and grandfather of Noah, is said by the Bible to have lived for 969 years, making him the greatest of the patriarchs and the oldest man in history. It therefore seems appropriate to give the name Methuselah to a specimen of the oldest known non-clonal (reproducing sexually) tree in the world, the bristlecone pine. In 2018, this tree was estimated to be 4,849 years old. It would have germinated from a seed around 2831 BC, before the Egyptian pyramids were built. Its exact location is kept secret to prevent too many tourists compacting the soil around the roots, or damaging it by taking living material. However, it lives somewhere in Methuselah Grove of Inyo National Forest at an elevation of between 2,900 and 3,000 metres (9,500–9,850 feet) in the White Mountains of California.

Bristlecone pines – the common name derives from the spines on the female cone – are generally not very tall trees, up to 16–18 metres (52–59 feet) high, with thick, scaly bark and twisted trunks and contorted branches, many of them bare. It was scientist Edward Schulman who first recognized the great age of these pines by examining their tree rings. Dendrochronology, discovered by Schulman's mentor, A. E. Douglass, at the Lowell Observatory in Arizona, is a scientific dating technique based on the study of the annual growth rings of trees, which can then be used to determine the timing of climatic events and thus form a record of changes in the environment.

Schulman was carrying out fieldwork in the White Mountains in 1953 when he discovered the largest of all the bristlecone pines, just below the treeline

WONDERS OF THE WORLD

High up in the White Mountains of California, ancient bristlecone pines stand far apart from each other in the remote and almost silent landscape, with little else growing around. The twisted and gnarled trees, with more dead branches than live, are a charismatic presence.

at over 3,353 metres (11,000 feet) elevation. A squat specimen measuring only 12.5 metres (41 feet) tall, it has a massive, fluted body consisting of fused, multiple trunks with a total circumference of 11 metres (36 feet). A local ranger called it the 'Patriarch Tree' and the grove it grows in the 'Patriarch Grove'. The setting looks like a moonscape, with nothing else growing in the surrounding arid landscape except for other gnarled bristlecone pines, the limber pine (*Pinus flexilis*) and the Sierra juniper. With very cold conditions in winter, high temperatures in summer, strong winds, virtually no rainfall and a white, rocky dolomitic limestone soil that few plants can grow in, this is an incredibly harsh environment in which only the toughest plants are able to survive. But this tree does and has done so for a very long time.

Four years after Schulman's discovery of the Patriarch Tree, following further fieldwork and collection of core samples taken from more living trees in the area using a borer, Schulman and his assistant Maurice E. Cooley discovered the first tree over 4,000 years old. They named it Pine Alpha, one of seventeen pines in a grove that are perhaps all over 4,000 years old, including Methuselah. In fact Schulman sampled but never studied a bristlecone that is probably even older, at over 5,000 years old. Sadly, Schulman died suddenly of a heart attack aged 49 in 1958. His article describing his amazing discoveries was posthumously published in *National Geographic Magazine*.

One question frequently asked is how have these trees survived for so long and in such an extremely harsh environment? There are many theories. One is that because of the severe conditions they grow far apart from each other, which means there is no danger of a fire spreading and destroying them and there is little or no competition for resources from other species. Also, because of their short growing season they tend to produce dense, very durable, resinous wood, which is more resistant to fungal decay and insect attack. And even if an individual is damaged by a lightning strike, small strips of living bark, which Schulman called 'lifelines', are able to keep the tree alive.

There are two other closely related species of bristlecone pines: the Rocky Mountains bristlecone pine, *Pinus aristata*, growing in Colorado, New Mexico and Arizona, and the foxtail pine, *Pinus balfouriana*, which is endemic to California. Both are long-lived, but neither reach the age of the Great Basin bristlecone pine. Unfortunately, bristlecone pines face an uncertain future because of climate change, increasing temperatures, an introduced fungal disease and invasion from pine beetles, but these trees are true survivors.

Candlestick banksia

Banksia attenuata

On 29 April 1770, the first Europeans set foot on the continent of Australia. Led by James Cook, the party included artist Sydney Parkinson and the indefatigable naturalists Joseph Banks and Daniel Solander, who wasted no time in starting to collect and record the astonishing flora of this new land. Because of the vast quantity of specimens they brought back on board the *Endeavour*, Cook named their first landing place Botany Bay.

Each spring tall, upright, vivid yellow inflorescences appear ... which sit on the tree like candlesticks on an old-fashioned Christmas tree.

Following the ship's triumphant return to Britain, the mission of describing and scientifically naming all the collected specimens began. One of the new genera of plants was named *Banksia*, in Banks' honour. Banks had collected a total of five different banksias on that voyage, but today we know that there are around 170 species of these unusual plants, which range in size from small shrubs to 30-metre (98-foot) tall trees. Characteristic of the dry shrublands and forests of Australia, banksias are an integral part of the ecology of these habitats. The nectar offered by their flowers is a valuable food source for many insects, birds, invertebrates and mammals, which means in turn that the banksias have no shortage of pollinators.

Banksia attenuata, known as the candlestick or slender banksia, is native to southwest Western Australia, on the other side of the continent from where Banks first saw the plants that would be named after him. This species was discovered some thirty years later by eminent Scottish botanist Robert Brown, who explored Australia with Captain Matthew Flinders (at the invitation of Joseph Banks) between 1801 and 1805, along with Kew gardener Peter Good and eminent botanical artist Ferdinand Bauer. Brown's work on Australian

The candlestick banksia produces tall slender spikes of nectar-rich flowers, which attract a variety of hungry pollinators. The flowers are followed by strange-looking woody seedpods.

WONDERS OF THE WORLD

plants proved to be extremely important, and he described and named around 1,200 species.

This particular species of banksia is a beautiful representative of the large genus. Growing to around 10 metres (33 feet) high, it characteristically has a wavy or undulating main trunk covered with a thick, orange-grey bark. Its pretty grey-green leaves are slender and serrated, narrowing to a point (hence the species name *attenuata*, meaning thin or narrow), and are a pale grey underneath. But it is the flowers that this species is renowned for. Each spring tall, upright, vivid yellow inflorescences (spikes of smaller individual flowers) appear, which are held above the leaves and sit on the tree like candlesticks on an old-fashioned Christmas tree. These striking inflorescences can be as much 30 centimetres (12 inches) tall and open in sequence from the base upwards, offering a staged sequence of meals for a wide variety of hungry pollinators – from honey possums to honeyeater birds, as well as bees and ants. To the local Aboriginal people the tree is known as *piara* (or *biara*) and they make a refreshing drink from the flowers, which is also used to treat coughs and colds.

In Western Australia this is the most widely distributed banksia and a key element of the open, semi-arid eucalyptus woodlands. It has been estimated that the candlestick banksia can live for up to 300 years, and like many other species in such habitats it is highly resilient to fire. Not only can it survive bush fires by re-sprouting from special 'epicormic' dormant buds on its trunks and from an underground 'lignotuber', but its bizarre-looking seed pods (or infructescences) have also evolved to survive fires. These seed pods, or cones, which develop from the pollinated flower spikes, have prominent mouth-like openings called follicles that remain firmly closed until subjected to heat and smoke. This evolutionary trick means that the seeds are only released into an environment that provides an ash-rich seedbed in which to grow, where they can wait for rains to fuel their growth. Cockatoos have been known to carry off the seed pods, and are thought to help in dispersing the seeds.

The region where candlestick banksias are found is a known biodiversity hotspot. It has a Mediterranean-like climate but is becoming increasingly drier and hotter, and fires are becoming more unpredictable and destructive. Studies have shown that *Banksia attenuata* has an evolutionary history of around 19 million years, and must have been able to adapt to shifts in the climate and environment in the past, and so it is to be hoped that this iconic species will also withstand future changes.

The handkerchief tree

Davidia involucrata

═══════

One of the most exciting botanical discoveries of the late nineteenth and early twentieth centuries, *Davidia involucrata* is a captivating and delightful tree in any garden setting. Imagine, then, the excitement of being among the first Western people to see it growing in its natural home, the rich temperate forests of the mountains of China. This was the privilege in 1869 of a French Vincentian missionary and keen naturalist and botanist, Father Armand David (Père David), who sent the first dried herbarium specimens to Paris. In 1871 it was declared a new species and named after him.

'The flowers and their attendant bracts ... resemble huge butterflies or small doves hovering amongst the trees.'

The challenge was then to obtain living plant material, including seeds, so that this most desirable tree could be introduced into cultivation. In 1899 Ernest Henry Wilson, one of the most successful plant collectors of the early twentieth century, was given a mission to locate Augustine Henry, chief medical officer and customs assistant in Shanghai. Henry had travelled to Yichang in Hubei province to obtain information on plants used in Chinese medicine, and had sent more than 15,000 dried herbarium specimens back to Kew for research and display, including *Davidia involucrata*. Henry pleaded with Sir William Thiselton Dyer, director at the Royal Botanic Gardens, Kew, and Sir Harry Veitch of the nursery firm James Veitch and Sons, for someone to be sent out to China to collect material of the many elegant trees he was encountering.

Wilson was chosen for the task as he was an up-and-coming, fit young horticulturist with a good eye for horticultural plants, though he had never been abroad and spoke no Chinese. Once he had reached China, Wilson first had to find Augustine Henry, who then drew a

BELOW LEFT A young healthy
specimen of *Davidia* ready
for planting. Ernest Wilson
had the privilege of being one
of the first Westerners to see
this tree in its native habitat
on his expedition to China.

BELOW RIGHT The hard-
coated seeds can take over two
years to germinate, so patience
is needed if trying to grow
this beautiful garden tree
from seed.

small map, about the size of a beer mat, showing the exact location of
the *Davidia* tree he had obtained the dried specimen from for Kew.
Wilson set out, undeterred by the tiny map, making the long and dangerous trek to the Yangtze Yichang gorges. In the autumn of 1899 he
eventually arrived at the spot that Henry had marked on the map, only
to find a stump. The tree had been cut down for wood to build a house
in the village. Henry's disappointment was only too evident in his
letters to Thiselton Dyer, but he stayed on in the area and in the following spring of 1900 found numerous specimens of *Davidia involucrata*
gracing the forests of the surrounding mountains. In November of that
year he collected a large quantity of seed, which he sent back to the
Veitch nursery for sowing.

Wilson wrote that *Davidia* was an aristocrat of the garden and
'the most interesting and most beautiful of all trees which grow in the
north temperate regions', saying that 'the flowers and their attendant
bracts ... resemble huge butterflies or small doves hovering amongst
the trees'. Most people would immediately agree with this description
when they see the spectacular hanging white bracts, which surround
the flower in late spring, fluttering in the slightest breeze, giving this

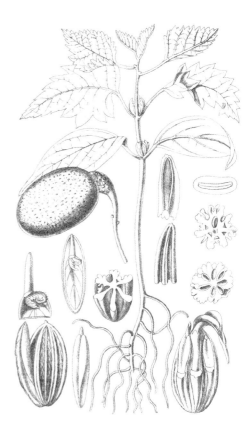

It is easy to see how *Davidia involucrata* acquired the two common names that we use today, the handkerchief tree and the dove tree. The pair of large white bracts that surround the small flowers flutter in the slightest breeze.

tree its two common names today – the handkerchief or dove tree. A deciduous tree, and the only species in its genus, *Davidia involucrata* grows up to 20 metres (66 feet) tall, with flaking bark and vivid green heart-shaped leaves. The small, spherical reddish flowers develop into dark green nuts containing six to ten seeds.

But who would be the first to germinate seeds successfully and introduce this tree to our gardens? The first batch of seed from Wilson's collection apparently failed to germinate and the seeds were discarded on the compost heap. A year later the heap was covered in young *Davidia* plants, as the hard-coated seeds can take over two years to germinate. The first flowers appeared in 1911. However, Wilson was unaware that another Frenchman, Père Farges, had also collected seeds of *Davidia* in 1897 and had sent them to Maurice de Vilmorin for his arboretum in France. A single one germinated and bloomed in 1906. So although Wilson's seeds were not the first to flower, he was responsible for the wider distribution of this glorious tree. Our gardens today would be so much poorer were it not for the early plant collectors who risked life and limb to find and collect such botanical treasures.

Coco de mer

Lodoicea maldivica

———

Endemic to just two tiny islands in the Seychelles, the extraordinary coco de mer palm can only be found in the wild in a handful of locations on Praslin and Curieuse. The palms can grow to between 25 and 50 metres (82–165 feet) tall, with huge, pleated, fan-shaped leaves up to 10 metres (33 feet) long. Its record-breaking fruits can weigh up to 40 kilograms (88 pounds) and measure half a metre across, and these contain the world's largest and heaviest seed.

These curious palms were once the stuff of myth and legend. Sailors believed they grew under water at the bottom of the Indian Ocean, and it was thought that male trees uprooted themselves on stormy nights and walked to find female trees, embracing them to pollinate their large flowers. People unlucky enough to witness such an event might go blind or even die. 'Coco de mer' is French for coconut of the sea and the name possibly arose from people having seen the huge seeds washed up on beaches or floating in the surf, although the fruit is heavier than water and seeds may float only when completely dried and empty. Also known as double coconuts, their rarity and suggestive rounded shape once made the seeds highly sought after and a valuable collector's item. Royalty and nobility prized them in their cabinets of curiosities, often elaborately mounting them in gold. Trade in the seeds is now closely controlled and they can only be sold with a permit, although illegal collecting is a problem.

Although the coco de mer has been extensively studied, it still guards some mysteries and secrets. It is thought that the sheer size of the fruit and the seed it contains is a result of the geographic isolation of this palm – known as island gigantism. Normally a parent plant's strategy would be for its seeds to be dispersed a fair distance away, by the wind or an animal, so that offspring are not in direct competition for light and nutrients. But these seeds are so heavy that they fall close

The record-breaking fruits of the coco de mer contain the world's heaviest seeds. These were once desirable collector's items. In the 1750s one is said to have sold for £400 – almost £70,000 today.

F.70

COCO DE MER

191

LODOICEA SECHELLARUM *Labill.*

Des deux pieds représentés sur le premier plan, celui de gauche est
un pied femelle, et celui de droite un pied mâle.

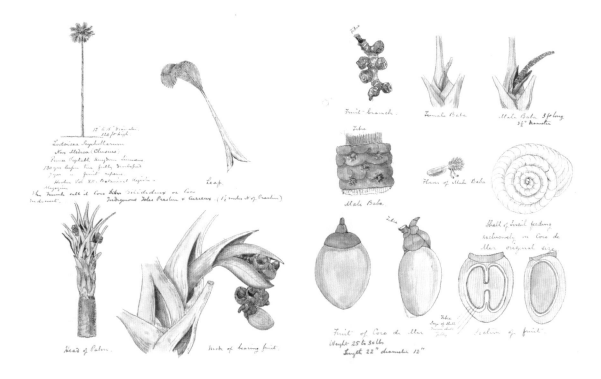

The sketch includes the following handwritten annotations:

Head of Palm.

Leaf.

Mode of bearing fruit.

Fruit-branch.

Female Baba.

Male Baba 3 ft long. 3½" diameter

Flower of Male Baba

Male Baba

Fruit of Coco de Mer Weight 25 to 30 lbs Length 22" diameter 12"

Shell of Snail feeding exclusively on Coco de Mer original size

Section of fruit.

ABOVE The pollination mechanism and ecology of this species are yet to be scientifically studied in detail, so it is still not known exactly how the flowers are pollinated in order to give rise to the enormous fruits. These sketches by nineteenth-century explorer and naturalist William Burchell show details of the anatomy of the coco de mer.

OPPOSITE The coco de mer palm can now be found in the wild in just a handful of locations in the Seychelles. Individuals can grow to between 25 and 50 metres (82–165 feet) tall, with male trees having 'catkins' 1.5 metres (5 feet) long.

by and grow in the shade of their parent. In fact the soil here also contains more nutrition than further away, as the fan-shaped leaves of the mature palm are extremely effective at channelling water and accompanying nutrients down the trunk to the soil at the base. As a result, stands of coco de mer occur and they are usually the dominant species in the forests where they grow.

It is still not understood exactly how the pollen from the 1.5-metre (5-foot) long 'catkins' of male trees is transferred to the flowers of the female trees, the largest female inflorescence of any palm. Some believe bees are the agents, others think lizards may be involved. The seeds can take six or seven years to ripen before falling, and then more time elapses before the cotyledon – the sprouting shoot – emerges. At around 4 metres (13 feet) in length this is longest known. This rope-like shoot, fed by the nutrient-packed seed, helps the new plant find the best spot to set down its roots.

This extraordinary palm is now under threat from harvesting, fires, introduced pests and human development. Although palms have been planted on a few islands near those where it grows naturally, the total population consists of only around 8,000 individuals, making this botanical phenomenon an endangered species.

Dawn redwood

Metasequoia glyptostroboides

At first sight this tree has perhaps one of the most formidable and unpronounceable botanical names. But there is an intriguing story behind its discovery and naming, involving ancient fossils and more recent political events. *Metasequoia* means 'like or close to a sequoia' and *glyptostroboides* is derived from the name of the similar-looking deciduous Chinese water cypress, *Glyptostrobus pensilis*. Even then the story has only just begun.

Metasequoia glyptostroboides, commonly known as the dawn redwood, is a deciduous conifer native to Hubei province in western China. It was originally described as *Metasequoia* in 1941 by a Japanese palaeo-botanist, Shigeru Miki, not from a living tree but from a 5-million-year-old fossil from the Pliocene epoch, which extended from 5.33 million to 2.58 million years ago. At the time of its identification, the tree was thought to have become extinct around 1.5 million years ago, and had once grown throughout the northern hemisphere at the same time as the dinosaurs roamed the planet. This is where things gets even more complicated.

In the same year, 1941, unaware of Miki's findings, it seems a Chinese forester named T. Kan (or Gan Duo), of the Department of Forestry of the National Central University in Nanjing, was carrying out a survey in Hubei province and came across a large, unidentified tree growing on the edge of a small village called Modaoxi, known today as Moudao. The tree, which was estimated to be over 400 years old, had a sacred shrine at its base and was clearly revered by the local villagers for its age and character, and they called it *shuisha*, meaning water fir. But it was nearly three years later that a Chinese forestry official named Zhan, who was exploring in the area, collected

> *It was originally described as* Metasequoia *in 1941 ... not from a living tree, but from a 5-million-year-old fossil from the Pliocene epoch.*

An early drawing of the original sacred tree discovered in 1941 growing on the edge of the village of Modaoxi, China, with a small shrine at its base. Both the shrine and the tree are still there today, albeit surrounded by a concrete fence for protection.

DAWN REDWOOD

the first samples of branches, leaves, cones and seeds for investigation and identification.

The Second World War postponed further research until more samples were collected and sent to Professor Wan Chun Cheng, one of the most eminent Chinese botanists of the twentieth century, who discussed the mysterious tree with Professor Hu Hsien Hsu, another Chinese botanist and a pioneer of modern botany in China. Hu Hsien Hsu was familiar with the work of palaeobotanist Shigeru Miki and the tree that Professor T. Kan had originally seen in 1941, and he linked them both together. Finally, in 1948, *Metasequoia glyptostroboides* was identified and named as a living tree from the earlier fossil records; Professor Hu Hsien Hsu is credited with the discovery today.

In 1947 the Arnold Arboretum of Harvard University in Boston funded an expedition to collect seeds from the original tree in Modaoxi. The then director, Dr Elmer D. Merrill, distributed them to arboreta and scientific tree collections around the world for cultivation trials. Professor Ralph Chaney of the University of California, who made his own expedition to the area of China where the trees grew, is credited with coining the common English name dawn redwood.

Notwithstanding such a convoluted name and complicated story, the dawn redwood is a beautiful, fast-growing tree with a symmetrically balanced pyramidal crown. The trunk develops basal fluting and twists with maturity, showing off the orangey-brown stringy bark. A deciduous conifer, it has feathery, light green foliage which develops vivid ochre tints in autumn before the leaves drop, making it one of the most perfect and popular ornamental trees. Although only about 5,000 trees are still found growing naturally in a few small valleys

in Central China, making it an endangered species in the IUCN *Red List of Threatened Species*, it is now one of the most extensively planted tree species around the world in countries with a temperate climate, including the United States, Britain, Japan, Chile and New Zealand. All these tree plantings date from the late 1940s on. The original and living type specimen of the dawn redwood in Modaoxi is surrounded by a concrete fence and has a thinning crown, most probably caused by the ground becoming compacted around it. In 1996 it was measured as being 7.1 metres (23 feet) in circumference above the fluted basal flare, a typical attribute of this tree, and 34.65 metres (114 feet) tall.

In 1957, a man called Qingxi Li planted 100 seedlings of dawn redwood, which he had acquired from Nanjing Forestry College, wanting to improve the tree-scape of Pizhou in Jiangsu Province where he lived and worked. These trees grew well and he propagated more. Then in 1975 he began to plant the 'Pizhou Dawn Redwood Avenue', which today at 47 kilometres (29 miles) long and containing 1 million trees is regarded as the world's longest avenue, beating the Japanese cedar avenue of Nikko at 35.41 kilometres (22 miles) long, which was planted in 1625.

Douglas fir

Pseudotsuga menziesii

Another of the giant evergreen conifers of the Pacific Northwest Coast of America, the Douglas fir can reach heights of over 100 metres (330 feet), competing with the redwoods (p. 216) and Sitka spruce (p. 54) for the title of the world's tallest tree. It is also one of the most widely distributed trees in the western USA, ranging from British Columbia in the north to California in the south. Further inland into the Rocky Mountains or south of California into central Mexico, a variety named *Pseudotsuga menziesii* var. *glauca* occurs, which is a smaller tree with a distinctive blue foliage, while the bigcone Douglas fir, *Pseudotsuga macrocarpa*, has cones double the size of the usual Douglas. The genus *Pseudotsuga*, meaning 'false hemlock', was named by the leading French botanist Élie-Abel Carrière in 1867. In addition to the two North American species there are also two from Asia: *Pseudotsuga sinensis* from China and Taiwan, and *P. japonica* from Japan. Of all the trees whose names resonate with tales of exploration and adventure, the Douglas fir must have some of the most intriguing associations, involving two intrepid explorers and plant collectors of the eighteenth and nineteenth centuries.

Of all the trees whose names resonate with tales of exploration and adventure, the Douglas fir must have some of the most intriguing associations.

In 1791 Archibald Menzies, a Scottish naturalist, botanist and surgeon on HMS *Discovery* under the captaincy of Captain George Vancouver during the 'Vancouver Expedition', saw and documented this tree on Vancouver Island. So the specific epithet, *menziesii*, honours its European discoverer. On the same voyage Menzies also collected seeds of the monkey puzzle tree (p. 232). However, it was only thirty-six years later, in 1827, that seeds of Douglas fir were introduced into cultivation in Britain by David Douglas, a Scottish gardener and botanist

Douglas firs naturally grow in dense forests, where they self-prune their lower branches so the conical crown starts many metres above the ground. Trees growing in open habitats, especially younger specimens, retain their branches much closer to the ground.

The attractive female cones with three-pointed (tridentine) bracts protruding from between the cone scales and the needle-like leaves that have a citrus-like aroma when crushed make the Douglas fir one of the easiest conifers to identify.

from Scone in Perthshire (who died in mysterious circumstances by being gored to death by a wild bull in a pit trap on Hawaii in 1834). Despite not being a true fir, the common name of the tree celebrates this prolific plant collector. Douglas introduced over 200 new species to cultivation from North America, including the Sitka spruce (*Picea sitchensis*; p. 54), Ponderosa pine (*Pinus ponderosa*), sugar pine (*Pinus lambertiana*), grand fir (*Abies grandis*) and noble fir (*Abies procera*).

Douglas fir is one of the easiest conifers to identify, with dark, deeply furrowed and rugged bark, pendulous female cones with strange three-pointed bracts protruding from between each of the cone scales, and needle-like leaves that have an aromatic smell of citrus-pineapple when crushed. Mature old growth forest, the natural home of these trees, is an important habitat for the red tree vole (*Arborimus longicaudus*), which nests high up in the Douglas fir's gracefully drooping branches and feeds on its needles. This in turn, along with other small mammals, provides food for the endangered spotted owl (*Strix occidentalis*).

As well as being ecologically important, the Douglas fir is also economically valuable. Its prodigiously tall, straight and unbranched trunk makes it one of the world's most important timber trees, and today it is grown extensively as a plantation forestry crop. The knot-free timber is light-brown in colour with hints of red and yellow through the grain, and is strong and resistant to decay. It is a very versatile wood and used for a wide range of purposes including construction joinery, cladding, veneer and plywood. It is also ideal for flagpoles.

In 1959 at the Royal Botanic Gardens, Kew, a flagpole 68.58 metres (225 feet) tall, made from a single piece of Douglas fir, was erected by the 23rd Field Squadron of the Royal Engineers. The tree it came from, estimated to be 370 years old and weighing 37 tonnes, was harvested from Copper Canyon on Vancouver Island and was presented by the Provincial Government of British Columbia to mark the bicentenary of Kew and the centenary of the province of British Columbia. Unfortunately, after fifty years, the weather and woodpeckers had taken their toll, making it unsafe, and it was dismantled by skilled steeplejacks in 2009.

The Douglas fir is now widely planted beyond its native range, including in New Zealand. In several countries in Europe the tallest tree is now a Douglas – for instance France, Germany and Italy. And despite several rival contenders for the title, the tallest tree growing in Britain today, measuring 66.4 metres (217 feet), is a Douglas fir planted in the 1880s near Inverness in Scotland and known as Dughall Mor, which is Gaelic for 'tall dark stranger'.

Back at home on Vancouver Island in British Columbia, an enormous old-growth Douglas fir known as Big Lonely Doug stands alone in a logging clearing on the banks of the Gordon River near Port Renfrew. In the winter of 2011, a forester called Dennis Cronin was surveying trees for logging. Standing under this impressive tree, 66 metres (230 feet) tall, the height of a twenty-three-storey building, and around a thousand years old, he decided to give it a reprieve. Instead of fixing a bright orange felling tag to the trunk, he tied a green ribbon to a protruding root with the words 'Leave Tree' written on it. It is the second tallest tree in Canada, just behind the present-day champion Douglas fir, the Red Creek Fir, which grows in a nearby valley. Big Lonely Doug will continue to grow, but not alone, as it will distribute its seeds for the natural regeneration of the forest.

Kauri

Agathis australis

Towering high above the surrounding canopy of lush sub-tropical trees and tree ferns in Waipoua Forest National Park on the North Island of New Zealand stands Tane Mahuta, Lord of the Forest. Its enormous grey-blue columnar trunk alone is imposing, measuring a staggering 13.8 metres (over 45 feet) in girth and is 17.7 metres (58 feet) tall. The total height of this giant tree, with its impressive canopy, has been measured at 51.5 metres (170 feet), and it is thought to be around 2,000 years old. It is a colossus among its slender neighbours, and many thousands of visitors each month come on a pilgrimage to pay homage to this member of botanical royalty, the largest kauri tree in New Zealand.

Kauri resin is exuded naturally.... Over millennia deposits built up in the soil and were excessively mined by 'gum diggers'.

Kauri (*Agathis australis*) is a member of the monkey puzzle family (Araucariaceae), one of the oldest coniferous families, with a fossil record dating back to the Triassic period, over around 200 million years ago. The only member of its family native to New Zealand, the kauri is found nowhere else. Trees usually grow to around 30–40 metres (98–130 feet) tall, with unusual blue-grey flaking bark, which sheds in small plates, resulting in a mottled and highly textured appearance. After young trees grow through the surrounding canopy, the crowns spread out and can become immense, with adult leaves that are long, broad and leathery. As they grow, kauris shed their lower branches, but once they are mature their sturdy upper branches are home to many species. Epiphytic plants such as ferns, orchids and even shrubs have been found growing among the branches, while a rich diversity of lichen species can be found lower down on their trunks.

In the Māori creation story the god of the forests, Tane Mahuta, is the son of Ranginui, the sky father, and Papatuanuku, the earth

Kauri can grow to an immense size, towering over all other trees in the forest. They are host to many other species, from orchids and ferns to birds and mammals such as brush-tailed possums, bats and rodents.

mother; by separating his parents in their close embrace, it was he who allowed light into the world and created night and day. He clothed his mother with vegetation, and all trees are therefore considered Tane's children. The Māori greatly valued the kauri, second only to the tōtara (p. 246), and created war canoes from single trunks.

At the time Europeans reached New Zealand extensive kauri forests covered parts of the North Island, but because of their impressive size, straight trunks and strong, golden-coloured wood, trees were soon heavily logged and they are now only found in any numbers in national parks. As well as once being the nation's most important timber tree, kauri was also the source of another useful product – resin. The Māori prized it for use in natural medicines, and also as a fire starter as it ignites easily. They also used the soot produced by burning the resin for tattooing. The 'kauri gum' industry was extremely economically important in the nineteenth century, with harvested resin ('copal') exported for use in the manufacture of paints, varnishes and linoleum. The resin is exuded naturally and in generous quantities by the tree to heal wounds and prevent bacteria and fungi gaining entry. Over millennia deposits of the resin built up in the soil and were excessively mined by 'gum diggers'. Once these deposits were exhausted, the resin

began to be harvested directly from living trees by 'bleeding' them, but this was soon found to have a severe impact on the trees and was ceased. Today, any gum found is often used for creating jewelry and art.

Kauri trees are a keystone species – one that is vital for the health of the surrounding habitat. They are also considered a *taonga* – a natural treasure – by the Māori, yet deforestation across much of their native habitat to make way for agricultural land, as well as fires and logging, has meant this species has generally disappeared from across the range where it once grew. New threats come from introduced possums, which strip the leaves, and a fungal disease, kauri dieback, which means that measures are being taken to conserve special, much-visited kauri trees such as Tane Mahuta. Efforts are also being made to restore kauri forests and there are signs of natural secondary regeneration on abandoned farmland, so it is to be hoped that these impressive trees, including Tane Mahuta and his kindred, with their immense lifespans, will continue to reign over the forests of northern New Zealand.

Durian

Durio zibethinus

Known in its native Southeast Asia as the 'King of Fruits', and famed more widely as the world's most malodorous fruit, the durian has gained a notorious reputation. Because of its overpowering smell it is banned from several airlines, hotels and the public transport system in Singapore. Eating the ripe fruits has been likened to consuming custard in an open sewer. But while its smell may be evocative of sewage, the fruit's creamy flesh is regarded as a delicacy. Mark Twain when travelling in the region was told that 'if you could hold your nose until the fruit was in your mouth a sacred joy would suffuse you from head to foot'. The flavour has been variously described as reminiscent of caramel, almonds or bananas, while for the naturalist Alfred Russel Wallace it called to mind cream cheese, onion sauce and even sherry. Opinion is sharply divided – in some people the durian provokes a deep disgust, in others a high regard.

In their native forests they are a favourite with monkeys, wild pigs and elephants, which can track them down from over half a mile away through the trees thanks to their pungent odour.

Native to Borneo, Indonesia, Malaysia and possibly also Sumatra, wild durian trees can grow up to 40 metres (130 feet) tall. The very large, oval, thick-rinded fruits can weigh a hefty 3 to 8 kilograms (7 to 18 pounds) and are covered with ferociously sharp spines. These develop from beautiful yellow-white flowers with reflexed petals, which are held in clusters directly on the branches and trunk. As with the fruits, they too have a distinctly unpleasant odour, like sour milk, but this is a direct signal that they are ready to be pollinated. These pendant flowers offer plentiful nectar and pollen for their main pollinators – species of fruit bats, which visit at dusk and throughout the evening. Once the bats have done their job the petals fall and the fruit begins to form.

The durian tree can grow up to 40 metres (130 feet) tall, forming a conical shape. The leaves are glossy green above and a pale bronze underneath, and the large, heavy fruits are covered in spines.

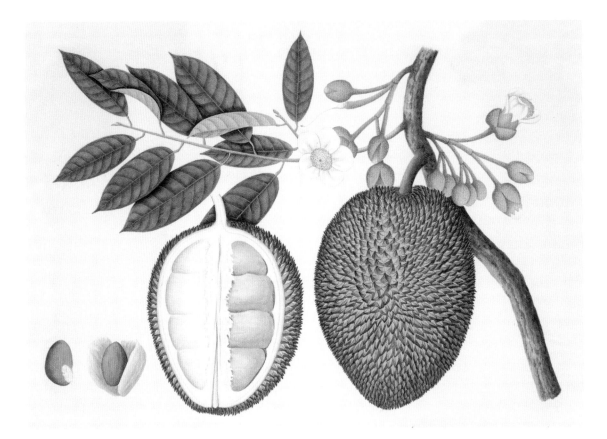

The yellow pulpy flesh of the durian fruit is said to taste exquisite, but smells so foul it is 'like eating custard in a sewer'.

When the fruits are ripe and fall they naturally split open. In their native forests they are a favourite with monkeys, wild pigs and elephants, which can track them down from over half a mile away through the trees thanks to the pungent odour they release. Having devoured the fruits, the animals then helpfully take the seeds away to be deposited elsewhere in the forest.

For their human admirers, durians are cultivated in many tropical Asian countries, including India, and also Australia. The valuable fruits are rich in vitamins and minerals and high in carbohydrates. There are over two hundred cultivars, each bred for a different taste and smell. People have their own favourites, and the variety known as 'Musang King' is becoming increasingly in demand in China. In 2012, two odourless and seedless varieties were introduced in Thailand called 'Longlaplae' and 'Linlaplae', which it is hoped will make the fruit even more popular and widely acceptable.

WONDERS OF THE WORLD

Red mangrove

Rhizophora mangle

Living in the ever-changing environment of the intertidal zone of the tropical coastline, by turns flooded by salt-laden seawater and then left exposed in baking mud, the red mangrove tree is a true survivor on the edge of what is possible for a flowering plant. Along with other species, this tree is one element of the group generally known as mangroves. It grows alongside the white mangrove (*Laguncularia racemosa*) and the black mangrove (*Avicennia germinans*) to form a tangled and specialized habitat fringing the western coast of Africa and tropical coasts of America, especially thriving along estuaries. These are some of the most productive and diverse ecosystems on Earth.

The red mangrove tree is a true survivor on the edge of what is possible for a flowering plant.

Like all mangrove species, the red mangrove has special adaptations to help it thrive in this challenging, waterlogged environment. Oxygen concentrations in the soil are extremely low, so the red mangrove grows aerial roots, pneumatophores, which can originate from the main trunk as much as 2 metres (6½ feet) and more above the ground. These allow oxygen to be absorbed through abundant pores (lenticels) in the bark and also act as stilts or props helping to anchor the tree. In addition, mangroves have adapted to tolerate high salt levels (they are halophytes) and can excrete salt from their roots.

Red mangroves are small evergreen trees, reaching up to 20 metres (65 feet) in height, with thick, leathery leaves. They produce their yellow-green bell-shaped flowers throughout the year, which may be either cross-pollinated by the wind or self-pollinated to produce red-brown berries. Unusually, these develop into large propagules, or seedlings, which can live attached to the parent for several months (known as vivipary) before detaching and floating away to find a suitable place to settle, where they quickly take root and grow.

T. 63.

Lacertus.

Rhizophora Mangle
Willd. sp. pl. 2 p 803

OPPOSITE Once pollinated, flowers of the red mangrove develop into long green propagules. These can then detach, float away and quickly develop into a new tree on a suitable shoreline.

BELOW Mangroves develop a tangle of aerial roots that help them 'breathe', but also support the plant in the intertidal zone and provide shelter for numerous creatures.

As well as being very unusual plants, mangroves are also extremely useful ones. A multitude of insects including ants and fireflies are at home in mangrove forests, while some amphibians, reptiles, birds and mammals, including bats and monkeys, visit and find food. The tangle of roots is home to barnacles and other molluscs, and when submerged the roots also act as a safe haven for young fish, crabs and shellfish, as well as larger species such as turtles and crocodiles. In their native range, mangroves are an integral part of a healthy coastal ecosystem and also support local livelihoods – from fishermen to those involved in tourism. They also provide timber and fuel, and their bark can be used to make rope and dyes. Mangroves are well known for protecting coastlines against tidal erosion and are also valuable carbon sinks.

Mangrove forests have suffered from habitat destruction, but many areas are being replanted in conservation initiatives involving local communities to stabilize and reclaim land. Such projects, which can be complex in their practicalities, often include the red mangrove (though it is now regarded as invasive in Hawaii). While it is hard to place a monetary value on trees, according to a recent WWF report the goods and services offered by mangrove forests are worth US $186 million per year to the world economy.

Quaking aspen

Populus tremuloides

A forest of quaking aspens is one of the most breathtaking sights to behold in North American forests in late autumn, when the leaves turn a rich, shining yellow and are highlighted against the smooth, greyish-white bark of the trunks. These stout, upright deciduous trees, which can reach up to 30 metres (100 feet) tall, are also affectionately known as 'Quakies'. It is not hard to understand where they get this common name from, as the diamond-shaped leaves, which are suspended on long stalks or petioles, flutter in the slightest breeze and produce a soft, whispering sound.

These trees are also affectionately known as 'Quakies' ... as the diamond-shaped leaves ... flutter in the slightest breeze and produce a soft, whispering sound.

Botanically, the quaking aspen is often mistaken for one of its closest relatives, the European aspen, *Populus tremula*, as it has many similarities and also has a wide natural distribution from western Europe to eastern Asia. In fact the species name *tremuloides* literally means having an appearance like *tremula*, the European aspen.

Quaking aspens can be found growing in moist, but not water-logged soils along the edges of coniferous forests, roadsides and areas cleared for forestry, and are regarded as the most widely distributed tree in North America. They occur as far north as Alaska and south into California, Arizona and Guanajuato in north-central Mexico, from Vancouver in the west to Maine in the east, and across every territory in Canada south of the tundra. In many places the quaking aspen makes up the major tree in forests. It has been the state tree of Utah since 2014, replacing the Colorado blue spruce (*Picea pungens*), which had held the honour since 1933, as the aspen makes up about 10 per cent of the total forest cover in the state compared to only 1 per cent consisting of blue spruce.

The leaves of aspens are held on long stalks, so they quiver or tremble even in light breezes, hence their botanical species name. Green in summer, the leaves turn a glorious shining yellow in autumn.

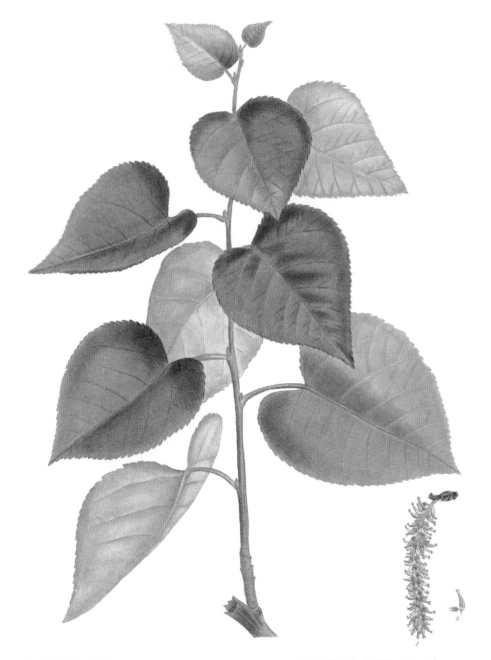

POPULUS Græca.　　　　　PEUPLIER d'Athènes. *pag.* 18¾

P. J. Redouté pinx.　　　　　　　　　　*Mixelle ainé Sculp.*

ABOVE The flowers of aspens are pendulous catkins produced in early spring before the leaves appear. Female catkins consist of strings of capsules, each of which contains about ten tiny seeds. These are easily dispersed on the wind when they mature in early summer.

OPPOSITE Aspens are found widely distributed across North America, often at high altitudes. They prefer moist sites and often grow in areas that have been cleared or disturbed, for instance by avalanches.

Although it is a flowering tree and produces viable seeds, *Populus tremuloides* rarely propagates sexually like most other tree species. Instead, it usually reproduces vegetatively from its roots and in this way forms large clonal groves, all sharing the same root system. This form of propagation is the reason why this species also holds the record as the heaviest known living organism. One 'tree' has 47,000 tree trunks and weighs around 6,000,000 kilograms (6,000 tonnes), with a root network occupying 43 hectares (106 acres). Estimated to be 80,000 years old, it was discovered by forest researcher Burton Verne Barnes in 1968, south of the Wasatch Mountains in Sevier County, south-central Utah. It is known as 'Pando', Latin for 'spread out', and also 'The Trembling Giant'. Incredibly, this is a single, male living organism with identical genetic markers.

Due to several factors, possibly including drought, the suppression of fires, which leads to more competition from other trees, and over-grazing by mule deer and cattle, Pando is now thought to be dying. Biologists are searching for the exact causes and trying to find an intervention that will save this prodigious tree. Above ground, the entire colony is made up of predominantly older tree trunks aged between 100 and 130 years old, the average lifespan for an aspen, which are reaching the end of their life. Very few youngsters are regenerating from suckers from the underground root system to fill the gaps left by the death of older stems. Where these young trees are successful, they are quickly grazed off by deer or trampled by cattle, leaving an age imbalance and a deteriorating colony. Selective ecological intervention using fire to prevent competition from the conifers that encroach into the outer edge of the colony has provided some space for new suckers to regenerate.

Like most other poplars, the white wood of aspens has relatively low strength and is useful mainly as pulpwood in the paper industry or for making packing crates and panel sheets including plywood. It does, however, provide a good source of food and dam-building materials for beavers, and the leaves of Quakies are eaten by a variety of moths and butterflies, birds and mammals, from rabbits to moose and bears.

QUAKING ASPEN

From a Photograph

SEQUOIA WELLINGTONIA

MARIPOSA GROVE, SOUTH CALIFORNIA

Redwoods

Sequoia sempervirens and *Sequoiadendron giganteum*

These two trees are true record breakers. They are often confused because both occur naturally in the states of California and Oregon in the northwest United States, and both are commonly called redwoods. The tallest recorded tree on the planet currently is a coast redwood, *Sequoia sempervirens*, but many tree experts believe that the tallest has not yet in fact been identified. The life of a tree is dynamic and constantly changing, and a challenger is probably out there waiting to be measured. However, the present champion, discovered in August 2006 by two naturalists, Michael Taylor and Chris Atkins, is a young tree named Hyperion, estimated to be around 600 years old. Growing in the Redwood National Park, California, it was measured as 115.9 metres (380 feet) tall by climbing to the top and dropping a measuring tape to the ground.

Coast redwoods are long-lived denizens of the forests and can reach ages of 1,000 to 1,200 years old, even up to 2,200 years, so Hyperion is still a mere sapling. There have been other contenders for the title of the world's tallest tree, and a mountain ash (*Eucalyptus regnans*; see p. 174) in Tasmania, at 114.3 metres (375 feet), came seriously close, but the measurements were not verified before it was cut down, so it has to take second place.

It is thought that when European settlers first arrived on the west coast of North America, the coast redwoods covered 648,000 to 769,000 hectares (1.6 to 1.9 million acres), but since then about 40,500 hectares (100,000 acres) have been lost to logging and forest clearance. These giants grow naturally in the state of California, from Monterey County in the south, to just 22.5 kilometres (14 miles) into Oregon in the north, in a long, narrow corridor, between 8 and 40 kilometres (5–25 miles) wide, along the Pacific Coast, known as the 'fogbelt'. The fogs that shroud these giants play an important role in the redwood

This is the giant redwood, *Sequoiadendron giganteum*, which grows on the western slopes of the Sierra Nevada in California and is the largest living organism on Earth. It is often confused with the coast redwood, *Sequoia sempervirens*, which is the tallest tree on Earth and grows along the Pacific Coastline.

OPPOSITE One of the scenes
from Edward Vischer's
Views of California of 1862,
of the Mammoth Tree Grove,
Calaveras County, California.
Just visible in the background
between the two tall standing
trees is one end of a felled
redwood.

BELOW The leaves of the
giant redwood are evergreen
and arranged spirally along
the shoots. The female cones
take eighteen to twenty
months to mature and remain
green and closed until the heat
of a fire opens them to release
an average of more than 200
seeds per cone.

ecosystem, helping to alleviate stress from drought by increasing relative humidity, lowering rates of evaporation and reducing transpiration, particularly in summer when the fogs are more frequent and rain scarce.

The first written description of the coast redwood appeared in 1769 in the diaries of a Franciscan missionary, Father Juan Crespí, who observed the trees while on the Spanish Portola Expedition near the Santa Cruz Mountains. He wrote: 'In this region, there is an abundance of these trees and because none of the expedition recognizes them, they are named red wood [*palo colorado*] from their colour.' Twenty-five years later, in the winter of 1794, Archibald Menzies of the Vancouver Expedition collected a sample which was used by botanist Aylmer Bourke Lambert to name the tree *Taxodium sempervirens*. The species name *sempervirens* means 'always' and 'green', thus distinguishing it from the other members of the genus *Taxodium* which are deciduous. Later, in 1847, the name was revised to *Sequoia sempervirens,* by Austrian botanist Stephen Endlicher. The name *Sequoia* came from the legendary Cherokee chief, Sequoyah, also known as George Gist, who is famed for inventing the Cherokee alphabet.

The giant redwood, *Sequoiadendron giganteum*, is another Californian record breaker. It grows on the western slopes of the Sierra Nevada, which are drier than the habitat of the coast redwood, and this redwood relies on natural fires to regenerate successfully. It has extremely thick spongy fire-resistant bark, often 45 centimetres (18 inches) thick and reddish-brown in colour, and its first branches are high up the trunk, out of the way of forest fires. The fires burn off the leaf litter on the forest floor, exposing the mineral soil that the seeds need to germinate in. The heat from the fires also opens the cones, which can remain unopened for up to twenty years, releasing the roughly 230 tiny seeds contained in each. These float down to the perfect seed bed below.

Giant redwoods can live for even longer than the coast redwoods – between 3,200 and 3,300 years. One of the oldest still standing, at a mere 2,500 years old, is also the most massive living organism on Earth. The incredible

General Sherman in Sequoia National Park, at 83.8 metres (275 feet) high but with a girth of 31.3 metres (103 feet) has a volume of 1,487 cubic metres (52,512 cubic feet) and is estimated to weigh 2,500 tonnes. Logging of these leviathans began soon after they were discovered in 1852 in the Calaveras Grove by bear-hunter Augustus T. Dowd. Casualties included the tree first seen, known as the Discovery Tree. However, the difficulty of transporting and sawing these huge trees, and their tendency to split or shatter when felled, meant that few made it from the groves to the saw mills. Most of the timber was wasted in the production of roofing shingles and fence posts, for which it was favoured because of its high durability. Entire felled trees were left on the ground in the forests, sparking action by conservation movements led by the Scottish-American naturalist and author John Muir, and others. Many redwood groves, including Yosemite Valley and Sequoia National Park, were then given the protected status that they so deserved. It is an eerie feeling to stand at the base of these venerable trees, listening to the creaking branches high above in the upper canopies, wondering what tales they could tell if only they could talk.

Podocarp, *Podocarpus*

THREATENED AND ENDANGERED

Although trees are the great survivors of the natural world, they are also under threat. Of the approximately 60,000 tree species in the world, it is thought that around 8,000 are endangered, though it is hard to be certain of numbers. Several appear elsewhere in this book, for example ebony, baobab, the kauri and dragon's blood trees. Many are directly important to us for food, medicine or timber, and all trees provide those unseen 'support services' such as controlling erosion and preventing flooding, and regulating climate. Every tree species is also a vital piece of a habitat's jigsaw, important to the wider ecology of the places where it grows.

There are myriad reasons why trees are threatened. Habitat loss and destruction are major factors – according to the World Wildlife Fund we are losing 18.7 million acres of forest every year through deforestation – while climate change, the movement of pests and diseases, invasive plants, pollution, over-collection and illegal logging all play their part. Two New Zealand species of great cultural significance to the Māori, pōhutukawa and tōtara, have been greatly reduced in numbers by a combination of factors, including the introduction of possums from Australia, which strip their leaves.

Perhaps surprisingly, some species considered under threat are relatively new to science. These include 'living fossils' such as the Chinese silver fir and the Wollemi pine. Once known only from the fossil record, they had been thought long extinct until rediscovered by chance in remote refuges. Another species was found by the side of a road by a schoolboy on his island home of Rodrigues. Only a single tree of the café marron was discovered and it took the skills and expertise of botanists at the Royal Botanic Gardens, Kew, to propagate it and secure its future.

Such endemics, which have evolved on remote islands and have nowhere else to go, are botanical Robinson Crusoes. On St Helena, in the wild Atlantic Ocean, the St Helena gumwood once covered vast tracts of the island, but is now reduced to a few small areas. Although not confined to an island, the beautiful Franklin tree was only known in one small area of Georgia in the United States, and has not been seen in the wild since the nineteenth century.

Even species that might be thought to be common or widely seen, such as the juniper and the monkey puzzle, are locally endangered. Despite their importance and usefulness to us, we still over-exploit trees and threaten their survival. Global regulations, along with research and science-based restoration projects, seed banks and botanic gardens, are all vital elements in conserving tree species around the world.

Juniper

Juniperus communis

A slow-growing and extremely hardy evergreen conifer, common juniper is often considered to be a shrub by many, but in the right growing conditions it can make a small, often twisted and contorted tree to 16 metres (52 feet) tall. It is distributed throughout the cool northern temperate regions of North America, Asia and Europe and into the Arctic Circle, with small relict populations in the Atlas Mountains of Africa, and has the largest geographical range of any other woody plant. A majority of plants have a low, spreading habit or are even prostrate, especially at higher elevations in exposed conditions, where trees are shaped by the wind. It is shade intolerant and often found growing in low-density woodlands, along with birch and pine, or on woodland edges and heathland.

Juniper's dense, very prickly foliage is a good deterrent to grazing animals such as deer, while at the same time supports many different birds as protective cover and nesting sites and as a source of food. The fruits, borne on female trees (there are separate male and female trees), are actually small cones containing two or three seeds. Green at first and berry-like, they take two to three years to ripen to a purple-black colour with a blue bloom. These berries are used in cooking for their pungent flavour and are the key ingredient that gives gin its unmistakable character. The word 'gin' is related to French *genièvre*, Italian *ginepro* and Dutch *jenever*, all of which are names for juniper. Legally the only plant ingredient, or 'botanical', that must be included in gin is juniper, and it is these berries that imbue the drink with its distinctive and invigorating pine-like taste, together with a harmony of other botanical ingredients including coriander seeds, citrus peel, angelica root and cinnamon. Just as a good wine is influenced by the grape, or a whisky is influenced by the oak cask, a good gin is influenced by the aroma and taste of juniper, which can vary according to where in

Juniper's green foliage is extremely prickly to deter grazing animals such as deer. The fruits, which are borne on female trees, are berry-like. They start out green and eventually, after two to three years, when the seeds inside mature, they turn a characteristic purple-black with a blue bloom.

THREATENED AND ENDANGERED

Berries ripening

Section of Berry
2 seeds in each.

Juniperus communis.
Southborough Common
Nov. 1897.

HERBARIUM KEWENSE.

Original drawing by the late
MRS. T. R. R. STEBBING.
Presented by E. C. WALLACE, ESQ.,
December, 1946.

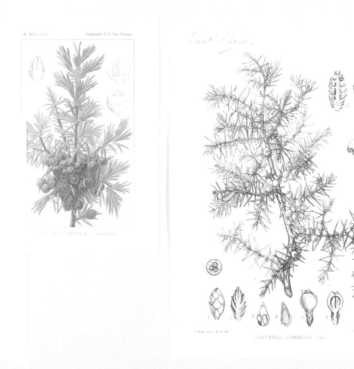

the world the juniper berries were collected. A fermented Finnish beer called Sahti is traditionally flavoured with juniper and hops and filtered through juniper twigs, producing a cloudy beer with a distinct taste said to be similar to bananas balanced by the bitterness from the juniper branches. Other juniper-flavoured drinks include a French fruity wine called 'genevrette', which is made from equal amounts of barley and juniper berries.

The wood of common juniper is relatively hard for a conifer, with tight, narrow annual growth rings and a golden-brown colour, but has little use apart from wood turning and carving by craftsmen, though pencils were once made from it. When burnt it produces an aromatic and almost invisible smoke, and there are tales that the wood of common juniper was used in the illicit stilling of whisky in the Scottish Highland glens to avoid attracting the interest of the local customs and excise man by the smoke. It is still used today to smoke various foods such as red meats and fish, giving them a unique, aromatic flavour.

There are many different species of juniper, and in various parts of the world they are associated with myth, magic and medicine. The earliest recorded medicinal use of the berries dates back to 1500 BC in Egypt in a recipe for curing tapeworm infestations, while the Romans

THREATENED AND ENDANGERED

OPPOSITE As well as being a key ingredient for flavouring gin, juniper berries, which are actually cones, have been used both medicinally and in cooking.

BELOW A fisherman sits beneath what is probably *Juniperus chinensis*. One of many different species of juniper, this is often used for bonsai. The watercolour and ink painting is by an unknown Chinese artist, commissioned by Robert Fortune on his final Chinese expedition in the mid-nineteenth century.

used juniper for purification and for stomach ailments. Nicholas Culpeper, the seventeenth-century English herbalist and physician, stated that juniper was almost without parallel for its virtues. In his *Complete Herbal* he recommended juniper berries for a wide variety of purposes including warding off venomous creatures and the treatment of flatulence. Today the berries are made into an oil which is high in flavonoid and polyphenol antioxidants and is said to be a powerful detoxifier and immune system booster.

Unfortunately, the common juniper has been slowly declining in the British Isles, especially in Scotland, mainly due to a fungus-like root rot disease (*Phytophthora austrocedri*). This pathogen attacks a tree's roots, killing the phloem which transports food around the plant, eventually girdling the main stem and causing death. However, seeds have been saved in Kew's Millennium Seed Bank, and because of this tree's vast geographic global range, it is not considered threatened at an international level.

Pōhutukawa

Metrosideros excelsa

Through November and December, in the lead up to Christmas – which is of course summer in the southern hemisphere – the pōhutukawa tree puts on the most striking display of crimson-red flowers. These can be so abundant and colourful that from a distance trees almost look as if they are on fire.

> *The flowers can be so abundant and colourful that from a distance trees almost look as if they are on fire.*

A member of the myrtle family (Myrtaceae), this beautiful evergreen is related to manuka and eucalyptus. Endemic to New Zealand, it grows mainly in coastal forests, where it flourishes in the warm dry conditions, but can withstand ocean winds and salt spray; the Māori name, pōhutukawa, means sprinkled with spray, referring to its preference to live by the coast. Mature trees can reach 20 metres (66 feet) in height, and are crowned with broad spreading domes of leathery foliage. The long leaves are hairy when young but change as they age – the upper surface becomes waxy while the underside remains covered in silvery hairs. This useful adaptation conserves water in dry conditions.

Pōhutukawa has great cultural significance for New Zealanders. In Māori mythology a young warrior called Tawhaki set out to avenge his father's death. He travelled to the heavens to ask for help, but fell to Earth and died. The red flowers of pōhutukawa represent his spilled blood. The Māori use the tree for boat-building, carvings, making a wide range of tools and in traditional medicine.

When European settlers arrived in the country and saw the tree flowering in December they gave it the English common name of New Zealand Christmas tree. It has since become an important symbol of Christmas traditions in the country. The tree's botanical names highlight its qualities and character: *Metrosideros* is derived from Greek

Pōhutukawa trees have immense significance for the Māori, and are considered *taonga* – treasured entities. They are also one of New Zealand's most iconic myrtle species.

The large leathery leaves and bright crimson flowers give this tree a very distinctive appearance. As it flowers in November and December, a common name is the New Zealand Christmas tree.

and means ironwood, as the slow-growing red-hued timber is heavy and hard; while *excelsa* is Latin for lofty or highest. The timber was used as firewood and also valued in shipbuilding for its resistance and durability.

Māori traditionally believe that when they die their spirit travels to a particular old sacred pōhutukawa tree (reputed to be over 600 years old) near Cape Reinga at the tip of the North Island. Here they travel down along the roots to a sacred cave and then on to the spirit world. The Cape is where the Tasman Sea meets the Pacific Ocean and is seen as the final stepping off point from this world.

Despite the reverence in which it is held, there are fears for this sacred tree, as well as other related myrtle species, as a result of the recent discovery of myrtle rust in New Zealand. This virulent wind-borne fungal pathogen, first seen in Brazil in the 1880s but only attracting attention in the 1970s, has recently spread quickly across

the Pacific and now threatens both endemic plants and important crops. Pōhutukawa are also under attack from invasive possums introduced from Australia, which love eating the leaves and buds, and also from human disturbance in the form of fires and soil compaction. It is estimated that by 1990 up to 90 per cent of the native coastal trees had been damaged or destroyed. Thankfully, since then many thousands of trees have been replanted and the future of this species is assured as seed banks have gathered abundant seed as a safeguard.

The Franklin tree

Franklinia alatamaha

This tree has the distinction of being named after one of the founding fathers of the United States. A member of the camellia family, Theaceae, which includes the tea plant, *Franklinia alatamaha* also has a remarkable story of discovery, introduction to cultivation and subsequent history. In 1765 King George III of Britain appointed John Bartram, a Pennsylvania Quaker farmer and plant collector, as the Royal Botanist for North America. This position allowed him to collect botanical herbarium specimens, seeds and living material to transplant to his own garden and to send overseas to European gardens, including the Royal Botanic Gardens, Kew. Later in the same year Bartram and his son William were exploring the banks of the River Altamaha in the state of Georgia, USA, when they came across a small grove of trees unknown to them. It was October, and the trees had fruit on them, but the Bartrams were not certain what the plant was. At first it was referred to as *Gordonia pubescens* as it had hairy fruits and was similar in appearance to *Gordonia lasianthus*, the loblolly bay.

> *It is a mystery why the Franklin tree is only known to have grown in such a restricted area, as William Bartram himself admitted.*

William Bartram later returned to the same spot and saw the trees in flower, describing them in his published *Travels* as 'very large ... of a snow white colour, and ornamented with a crown or tassel of gold coloured refulgent staminae'. He considered the tree to be 'of the first order for beauty and fragrance of blossoms' and collected material which was studied in London by Daniel Solander, a Swedish naturalist and student of Carl Linnaeus, who identified it as a new plant. The genus was called *Franklinia* in honour of John Bartram's close friend, Benjamin Franklin, the statesman and author who helped draft the American Declaration of Independence.

The species name came from the river (though in a variant spelling) where the grove of trees once grew – once grew, because this is the only place the tree has been observed growing naturally and it has not reliably been seen since 1803 and is considered to be extinct in the wild. William Bartram propagated some of the plant material he had collected and grew the tree in his garden in Pennsylvania, the first botanical garden in the USA, which John Bartram had established on the banks of the Schuylkill River near Philadelphia in 1728. All Franklin trees growing around the world today are descended from these trees planted in the Bartram garden.

Franklinia alatamaha is a large deciduous upright shrub or small tree, which in ideal conditions can grow to 10 metres (33 feet) high. In cultivation it more usually makes a modest tree around 4 to 7 metres (13–23 feet) tall. It freely suckers and can be grown with multiple trunks or as a single-stemmed tree, showing off the grey bark, which in mature specimens is marked with vertical, ridged striations. The elongated deep green leaves can reach 15 centimetres (6 inches) long, turning a rich orange-red in autumn. The Franklin tree is prized for its large, fragrant, white, camellia-like flowers produced in summer, with a scent reminiscent of orange blossom or honeysuckle. However, in cultivation in Europe the tree does not begin flowering until late in autumn, probably because summer temperatures are not as high as those in its natural home in the eastern United States, and this heat is needed to ripen the new soft wood that bears the flower buds. The spherical fruits that follow pollination can take over a year to mature. When ripe, the seed case splits open, releasing the viable seeds.

It is a mystery why the Franklin tree is only known to have grown in such a restricted area, as William Bartram himself admitted. It is equally uncertain why it disappeared from the wild, with possible causes including a pathogen washed down from cotton plantations, or destruction by people, flood or fire.

Although it can be difficult to cultivate or transplant, once established *Franklinia alatamaha* is long-lived and well worth the trouble and effort. Anyone who successfully grows this beautiful tree is also helping to conserve it for the future and prevent it becoming extinct.

Monkey puzzle

Araucaria araucana

Looking like the prehistoric relic that it is, this unmistakable, unusual tree has a tall straight trunk often topped in older specimens with a crown of tiers of horizontal branches which bristle with rows of stiff, spiny leaves. It is the 'monkey puzzler', an ornamental tree that was highly prized and widely planted as an individual specimen in parks and larger gardens by the Victorians. They were fascinated by its eccentric shape and form, which made it more of a curiosity than a useful plant in the garden. Belonging to the family Araucariaceae, which was living from around some 200 million years ago, its sharp, needle-like leaves would have provided protection from ancient grazing animals that have long been extinct, unlike the tree itself.

Like all good botanists, rather than eating the seeds Menzies pocketed them, and when back on board HMS Discovery *he planted them.*

Araucaria araucana grows naturally in forests in south-central Chile on volcanic soils on the western slopes of the Andes at elevations of between 900 and 1,800 metres (3,000–6,000 feet) and in the coastal cordillera in the Parque Nacional Nahuelbuta, with a small population in the Andes in southwest Argentina. In 1990, it was declared a Natural Monument and is the national tree of Chile.

The first Westerner to discover the tree was the Spanish explorer Don Francisco Dendariarena, in 1780, but it was introduced into cultivation fifteen years later in 1795 by Archibald Menzies, a surgeon, naturalist and botanist on HMS *Discovery* under Captain George Vancouver of the British Royal Navy during the 'Vancouver Expedition'. This expedition, which lasted four and a half years, had orders to carry out an accurate survey of the Pacific coast and navigable rivers into the interior of North America, while also gathering information and facilitating the research needs of Menzies. On the return leg of the epic voyage,

The monkey puzzle is a prehistoric-looking evergreen tree which can live for a thousand years and reach 50 metres (165 feet) tall. On young specimens in the garden the branches are retained right down to the ground and the tree only begins to shed them as it grows to maturity.

THREATENED AND ENDANGERED

MONKEY PUZZLE

Monkey puzzles have sharp, spiny leaves, which cluster in spirals around the branches. Male cones (left) and round female seed cones (right) are borne at the tips of the branches. Each female cone can hold up to 200 nut-like seeds that can take up to two years to mature.

Vancouver docked in the port of Valparaiso in Chile to carry out essential repairs and refitting of his ship for the long journey home. During the five weeks of their stay, Vancouver and Menzies were invited to dinner by the Viceroy of Valparaiso, at which they were served an unusual nut for dessert. Like all good botanists, rather than eating the seeds, Menzies pocketed them and when back on board HMS *Discovery* he planted them.

By the time the expedition returned to England, he had five young monkey puzzle saplings, which he presented to Sir Joseph Banks, garden adviser to King George III at Kew. The trees survived in the gardens there until 1892, where they were the focus of great botanical interest and curiosity. The nineteenth century also saw the genesis of the common English name in use today – the 'monkey puzzle tree'. The story goes that Sir William Molesworth of Pencarrow in Bodmin, Cornwall, was the proud owner of a young tree in his garden and was showing it to friends when one of them, a lawyer, remarked that it 'would puzzle any monkey to climb'. It thus became known as the 'monkey puzzler', despite there being no monkeys in Chile, or Cornwall, and the letter 'r' was later dropped.

James Veitch, of Veitch Nurseries of Exeter, had seen Menzies' trees in the gardens at Kew and employed a Cornish plant collector named William Lobb to find and bring back seeds of the monkey puzzle. Lobb collected over 3,000 seeds from cones that he obtained by shooting them off the branch tips at the tops of the trees, which were impossible to climb. His porters could then gather the nuts from the forest floor. By 1884 young trees were available for sale from the seeds Lobb had sent back to Veitch in 1842.

The monkey puzzle is a majestic evergreen tree which can live for a thousand years. It can reach up to 50 metres (165 feet) tall, with a trunk of large diameter covered in fire-resistant bark and a base resembling an elephant's foot. Young trees have branches down to the ground, but lower ones are shed as the tree grows. The sharp, shiny, reptilian leaves are spirally arranged around the horizontal branches, the ends of which bear the large round female seed cones holding up to 200 nut-like seeds, 3 to 4 centimetres (around 1½ inches) long. These can take two years to mature and ripen. The Mapuche people in the Araucania region, after which the scientific name is derived, believed that the monkey puzzle was sacred. They ate the seeds or piñones, which are rich in carbohydrates and protein, perhaps in a dish similar to the one served up to Menzies and Vancouver in Valparaiso.

The wood of the monkey puzzle is light and soft and was once used for building, flooring and paper pulp, and also for ships' masts, leading to logging and deforestation. But today international trade in the timber is illegal as the tree is listed under the CITES treaty and is on the IUCN Red List as an endangered species. Monkey puzzle trees can still be found in the Cordillera de Nahuelbuta, part of the Chilean Coast Range. The name Nahuelbuta derives from the local Mapuche language, with *nahuel* meaning jaguar and *futa* big, as these large cats live among the magnificent trees. On the wall of an old farmstead here can still be seen a plaque bearing the inscription 'Marianne North 1884'. North was a well-known Victorian botanical painter who travelled around the world twice over a period of thirteen years, painting rare plants in their natural habitats. She must have undertaken the journey to this place alone, most probably on a horse, with all her painting materials and supplies. She donated all her artworks to the Royal Botanic Gardens, Kew, six years before she died, and 832 of her paintings and 246 different types of wood, including *Araucaria araucana*, are displayed in a purpose-built gallery there.

Chinese silver fir

Cathaya argyrophylla

Jinfu Shan, or Golden Buddha Mountain, part of the Dalou range in the upper reaches of the Yangtze River, within the Nanchuan district of Chongqing in China, is home to a rare and beautiful conifer. On this limestone mountain, which stands at 2,251 metres (7,385 feet) high, is a small, closely guarded population of *Cathaya argyrophylla*, known colloquially as the Chinese silver fir or *yin shan* in Chinese. *Cathaya* is a reference to a historical name of its home country, China, and the specific epithet *argyrophylla* is Greek for 'silver leaf', describing a striking feature of the underside of the tree's needles.

This unusual and elusive tree, the only representative of its genus, was discovered as recently as 1955 by Chinese scientists in southeastern Sichuan.

This unusual and elusive tree, the only representative of its genus (making it monotypic), was discovered as recently as 1955 by Chinese scientists in southeastern Sichuan, and was identified as related to plant fossils dating back to the Pliocene. It has since been found in a few other isolated localities in Hunnan, Guangxi and Guizhou, and favours inaccessible open slopes and ridges on mountains at elevations of between 900 and 1,900 metres (2,950–6,235 feet) above sea level, often shrouded in thick cloud and with abundant moisture. The Chinese silver fir appears to be a naturally scarce species growing in mixed evergreen broadleaved forests – it is estimated that there are only between 500 and 1,000 mature individual specimens. It can reach 20 metres (66 feet) or more tall, with a straight, columnar trunk and dark grey, flaking bark. The needle-like leaves, around 4 to 6 centimetres (1½ to 2½ inches) long, are dark green above with a gleam of silver in two bands beneath, hence the name.

A 'living fossil', the Chinese silver fir can make an imposing tree to 20 metres (66 feet) tall with a columnar trunk, almost horizontal branches and dark grey flaking bark. It is rare in its native habitats in China, with trees now strictly protected in nature reserves.

Fortunately, the majority of the trees in China are now in nature reserves where they receive the highest levels of protection, with access tightly controlled. Very few Westerners have seen this 'living fossil' in

THREATENED AND ENDANGERED

237

its natural habitat. One botanical expedition to Jinfu Shan in 1996 had first to gain permission to visit the mountain from representatives from the local police, the local tourist bureau, the forestry department, the public security bureau, the Chinese army and the mayor's office, and even then was escorted by six Chinese officials. Every niche and foothold on a limestone outcrop beneath where the silver firs grew had been filled with concrete or chipped off smooth to prevent anyone gaining access to the trees, which loomed out of the mist above.

At that time collection of living plant material was restricted and there was a national embargo on the export or distribution of trees or seeds. The silver fir, reportedly referred to by Chinese botanists as 'the giant panda of the plant kingdom', thus remained largely unknown. Would this tree ever be found growing outside China? The official embargo was later lifted, and seed was distributed through forestry institutes and botanic gardens, including those in Sydney and Edinburgh, and the Arnold Arboretum of Harvard University. It is still a rare tree in cultivation, however, and as yet there are no mature trees growing elsewhere, but at least the population in China is receiving high-level protection, including a ban on logging imposed by the Chinese government. But natural regeneration seems to be poor for a number of reasons, and the tree's seeds are eaten by rodents, flying squirrels and the silver pheasant, which also feeds on seedlings.

The few cultivated trees in gardens, while still young, are now flowering and setting fertile seed for the first time. This has germinated and more young trees will soon be added to the established collections, further ensuring the survival of this threatened tree. *Cathaya argyrophylla* is not as widely known as an ornamental in the garden as it deserves, but wherever it is grown it will certainly be cherished for its beauty and its rarity.

The needle-like leaves of *Cathaya argyrophylla* are dark green above with a gleam of silver in two bands on the underside, hence both the specific epithet, *argyrophylla*, meaning 'silver leaf' in Greek, and the common name in English.

Wollemi pine

Wollemia nobilis

The oldest known related fossil to the Wollemi pine dates from 90 million years ago, and the tree was thought to be extinct until a few specimens were discovered growing in remote canyons not far from Sydney in Australia in 1994.

On 10 September 1994, David Noble of New South Wales National Parks was bushwalking alone in the remote and undisturbed steep-sided sandstone gorges of the Wollemi National Park in the Blue Mountains, only about 150 kilometres (just over 90 miles) northwest of Australia's largest city, Sydney. He came across an unfamiliar, very unusual-looking tree that he had not seen before during his many hikes in these wild canyons. Having collected a small sample of foliage, he took it back to the Royal Botanic Gardens in Sydney to be identified by the garden's taxonomists. It was a dramatic discovery that would astound the plant world, as it was recognized as a new tree species unknown to science. The tree was later named *Wollemia nobilis*, the Wollemi pine. *Wollemia*, the genus, is named after the Wollemi National Park, with the Aboriginal word 'wollemi' meaning 'look around you, keep your eyes open and watch out'; the species name *nobilis* reflects both the tree's qualities and its discoverer, David Noble.

The Wollemi pine is not a true pine at all and is a member of the very primitive family Araucariaceae, which was once abundant in the world's forests during the Jurassic and Cretaceous periods some 200 million to 65 million years ago. The other two genera represented in this family today are now mostly restricted to the southern hemisphere – the kauri, *Agathis* (p. 202), and the monkey puzzle, *Araucaria* (p. 232). The oldest known related fossil to the Wollemi pine dates from 90 million years ago and it had been presumed that the tree had become extinct some two million years ago. As it was only

THREATENED AND ENDANGERED

The Wollemi pine is a member of the same plant family as the monkey puzzle, and like its cousin its leaves are flattened and arranged spirally. The round female cones are also similar in appearance to those of the monkey puzzle, but are much smaller in diameter.

known from the fossil record before its dramatic reappearance in 1994, it is a 'Lazarus' taxon or 'living fossil'.

Wollemia nobilis is a tall evergreen coniferous tree reaching up to 40 metres (130 feet) in height, with a trunk diameter of around 1.2 metres (4 feet). On maturing trunks over ten years old the bark is distinctly knobbly and has been described as resembling bubbling chocolate. The tree's unique branching habit means that it never produces lateral branches from the main framework of branches growing off the trunk. Like its cousin the monkey puzzle, the foliage is arranged spirally and the leaves are flattened in two or four ranks, making it easy to identify. During dormancy in winter, the terminal buds are covered in a white resin known as a 'polar cap', which protects the growing tips from damage by cold temperatures. This is thought to be one reason why this tree survived the ice ages. Once spring arrives the fresh, soft, lime-green foliage breaks through the cap and begins to grow, gradually becoming a mature blue-green.

Two more small groves of Wollemi pine have been located since the original discovery, though fewer than a hundred mature specimens remain. Most have several multiple trunks growing from the base, with some trees having up to a hundred. This system of natural coppicing may have evolved as a defence against fire and rock falls in the steep gorges where the tree naturally grows, also ensuring its survival into the present day. However, it is also possible therefore that the trees are clonal, and it has been shown that there is very little genetic variation between individuals.

Classified as critically endangered, the Wollemi pine is now protected in Australia and anyone found venturing into the remote canyon, the exact location of which is kept secret, will be prosecuted. This penalty has been implemented to prevent the introduction of a plant disease, an aggressive water mould (*Phytophthora cinnamomi*), which can cause huge environmental damage to fragile plant populations when introduced on people's footwear. As part of the conservation strategy for the tree, young plants have been cultivated and distributed or sold around the world. In the Mount Tomah Botanic Gardens in the Blue Mountains an established planting of Wollemi pines grows in a fenced valley that mimics the trees' natural home and safeguards the gene pool of this botanically fascinating tree. The Wollemi pine can rightly be called the dinosaur of the tree world and a living link to an ancient time. Perhaps if we had followed the Aboriginal meaning of the word wollemi, we might have rediscovered it a lot sooner.

St Helena gumwood

Commidendrum robustum

A gnarled, tenacious-looking tree, this gumwood is the national tree of St Helena, a remote dot in the ocean, 1,930 kilometres (1,200 miles) off the coast of Africa. Native to this windswept volcanic South Atlantic island, it grows nowhere else on Earth and once formed forests that covered entire hillsides over half of St Helena. However, through deforestation, agriculture, the introduction of non-native species and the tree's usefulness as timber for British settlers who arrived in the seventeenth century, the St Helena gumwood has been reduced to two small wild populations, isolated from each other. It is now classed as critically endangered.

> *Native to this windswept, volcanic South Atlantic island, it grows nowhere else on Earth, and once formed forests that covered entire hillsides over half of St Helena.*

The St Helena gumwood, with thick and hairy grey-green leaves forming an umbrella-like canopy, grows up to 8 metres (26 feet) tall and produces small white pendulous flowers that hang from the ends of the stems. These unusual-looking flowers are visited by a range of insects, including an endemic hoverfly, and once pollinated they develop into simple fruits that each contain a single seed. The seeds are dispersed by the wind and readily germinate and grow if they land in an ideal spot and the seedlings escape being eaten by grazing animals.

In 2013, scientists counted just 679 individual wild trees, which were struggling to produce a new generation of saplings because of browsing by livestock, rats and rabbits, as well as invasive weeds that smothered seedlings. Mature trees were also suffering from invasive pests such as the sap-sucking jacaranda bug. However, this distinctive tree is now protected and has become a focus for a concerted conservation programme by the islanders. 'Gumwood Guardians' are planting new saplings, helping to remove weeds and rats from the areas where

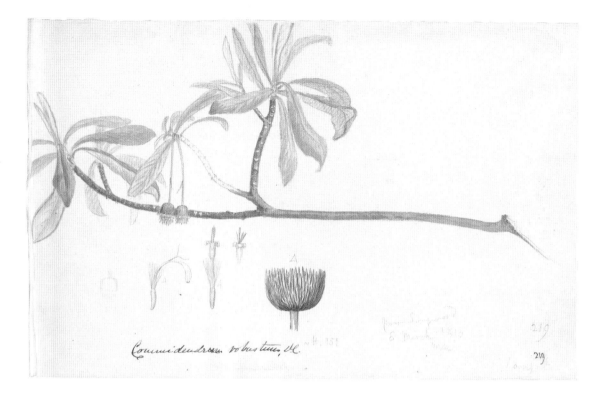

Comuidendron robustum, DC. [handwritten annotations]

This sketch by traveller and keen plantsman William Burchell, dated March 1810, accurately depicts the leaves and pendulous white flowers of this unusual tree, which was already declining by the time of Burchell's visit.

the trees naturally grow, and erecting fencing to prevent browsing by cattle and goats.

The Millennium Forest project has been a particularly important initiative, led by the St Helena National Trust and supported by the resident community. Around 10,000 trees across 35 hectares (86 acres) have been planted since 2000, repopulating a site once known as the Great Wood, which was entirely destroyed by early settlers who arrived in 1659. In total around 250 hectares (618 acres) have been set aside for the reforestation of this gumwood and other endangered trees, and it is estimated that another 55,000 trees need to be planted. The Trust, aided by volunteers, the St Helena government and the Royal Botanic Gardens, Kew, is also striving to grow and replant many other endemic species of the island. Seeds are being banked to ensure their survival and an online herbarium created to aid research.

The history of the development and destruction of the flora of St Helena is typical of many islands, but the loss of species which have evolved over millions of years is not inevitable. This is another positive example of how committed and enthusiastic people can try to reverse the process and ensure the survival of such special trees.

Café marron

Ramosmania rodriguesii

If ever proof were needed that an inspiring teacher can change the world, then the story of the café marron tree provides it. This species, found only on the tiny island of Rodrigues in the Mascarene Archipelago of the Indian Ocean, had been believed to be extinct for many decades as it had not turned up in botanical surveys. However, in 1984, schoolteacher Raymond A-Keeh encouraged his pupils to head out into the local area to collect plants and see if they could find any native species. One schoolboy, Hedley Manan, surprised everyone by returning with a sample that was later identified as coming from a single surviving specimen of the supposedly extinct café marron.

This charismatic small tree, just 2–4 metres (6½–13 feet) tall, differs markedly in appearance in its juvenile and adult forms (an example of heterophylly). The young plant has linear, strap-like leaves, up to 30 centimetres (12 inches) long, which are marked with small blotches of dark brown, black and crimson and have a pinkish stripe down the centre. This colouration is thought to have evolved to deter grazing animals, including possibly a giant land tortoise and the dodo-like Rodrigues solitaire, both now extinct. When the plants reach a height of around 1–1.5 metres (3¼–5 feet), the foliage is transformed into pairs of shorter, glossy dark-green oval leaves. Beautiful white star-like flowers are produced in inflorescences on male plants and singly on female plants.

The sole surviving wild tree was growing close to a road and was popular in local medicine as a cure for hangovers, so it was quickly protected by fences. In 1986, cuttings were flown to the Royal Botanic Gardens, Kew, and were taken to the specialist nursery and the micro-propagation unit. One cutting was successfully rooted and grown on in the nursery, and by taking successive cuttings a small healthy population of clones was soon growing. Eventually, these began to flower,

The oval leaves of the mature café marron tree are deep-green and glossy – a perfect foil for the white star-like blooms of this endangered tree. The dried specimen of the leaves is held in the herbarium of the Royal Botanic Gardens, Kew.

THREATENED AND ENDANGERED

but for many years staff at Kew could not work out how to pollinate the flowers and therefore – crucially for the tree's future – produce viable seed. That was until Carlos Magdalena became involved. Carlos, a dedicated horticulturist at Kew, now has a reputation for saving plant species that are considered the 'living dead', with no real prospect of survival in the wild. Because of his many successes with endangered species he has even been called the 'Plant Messiah'.

In 2003, Carlos began to investigate what conditions were necessary for the café marron to be pollinated and set seed successfully. He discovered that a hotter, brighter environment was the key, and with dedicated hand-pollination he eventually had fruits developing with seeds inside. Healthy seedlings were germinated and the distinctive foliage of the young plants became apparent, and also later the difference in flowering habit of the males and females. Cross-pollination of male and female flowers produced a new, much healthier generation. Around fifty saplings were carefully raised and many have now been successfully repatriated to Rodrigues. The story of the café marron highlights the vulnerability of island plants, but it is hoped this tree has now been rescued from the brink of extinction and may once more grow freely in its small native home.

Tōtara

Podocarpus totara

Podocarps belong to the large ancient coniferous plant family known as Podocarpaceae, and can date their ancestry back to the supercontinent of Gondwana, which dominated the southern hemisphere of the planet millions of years ago. As Gondwana split and slowly drifted apart, the podocarps, along with other species, began to change and evolve, and today members of this family can still be found growing across the southern continents. New Zealand has thirteen species of the trees called podocarps, with some of the best known referred to by their Māori names of *rimu*, *kahikatea*, *miro*, *mataī* and *tōtara*. The tōtara are actually themselves a collection of four different species of the botanical genus *Podocarpus*. Together, the podocarps once grew in vast pristine forests across the islands.

Tōtara became closely linked with Māori culture, and the wood was used extensively for their beautiful, elaborately ornamented carvings.

One of the most valued of these species is *Podocarpus totara*. Endemic to New Zealand, it grows wild on the North Island and parts of the South Island. Towering up to 40 metres (130 feet) tall in the fertile lowlands, this conifer has a distinctive appearance, with linear flat leaves, which are stiff and leathery, and thick, shaggy, reddish bark, which peels away in strips so that the trunks are ridged and furrowed. This species has separate male and female plants (it is dioecious), with the males producing pollen-bearing catkins and the females bearing red and green fleshy cones which look a little like berries and are edible.

The uses of this slow-growing forest giant are many. Its high-quality timber is prized by the Māori above all others, including the kauri (p. 202). It is light, strong, extremely durable and resistant to rot; it is also easily worked and its reddish-brown hue gives a pleasing, tactile quality to anything made from it. The wood was used to

This ancient endemic conifer can grow up to 40 metres (130 feet) tall and is a key part of the natural forests of New Zealand.

THREATENED AND ENDANGERED

OPPOSITE The wood of the tōtara is easily worked and was used by the Māori for their carvings, including masks, as seen in the centre of this drawing of Māori and other objects by Sydney Parkinson, an artist on James Cook's *Endeavour* voyage.

BELOW Special war canoes or *waka*, the largest of which could hold up to a hundred people, were made from tōtara heartwood. In this lithograph by Augustus Earle, a Māori chief addresses a gathering of warriors from a canoe pulled up on the beach.

construct houses, furniture, tools, weapons and instruments but most especially canoes (*waka*). Strong tōtara heartwood was particularly favoured for *waka*, which were made in a variety of sizes. The largest, holding up to a hundred people, were special war canoes that carried the fiercest warriors to battle. Tōtara became closely linked with Māori culture, and the wood was used extensively for their beautiful, elaborately ornamented carvings, which still today are used to tell stories about their ancestors and history and in some cases reputedly offer protection. Māori carving developed a very distinctive character over time, evolving from its Polynesian origins, and different carving styles arose in different areas of the country. Carvers became highly respected members of Māori communities and the art is firmly associated with tribal identities today.

Traditional Māori medicinal uses of tōtara included burning the bark to produce smoke to treat skin complaints and venereal disease, making an infusion from the leaves for upset stomachs, and boiling the inner bark to create a tonic for fevers. The heartwood of *Podocarpus* contains a chemical compound called totarol, which is responsible for the timber's resistance to rot. This is of potential interest as research has shown it to have anti-bacterial properties and it may have a use in medicine as well as in cosmetics and even dental products.

THREATENED AND ENDANGERED

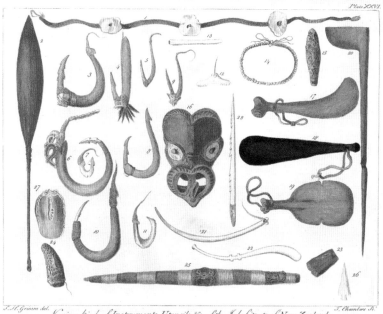

Podocarps were once extensively logged by the European settlers of New Zealand, who valued the wood for many of the same purposes as the Māori, but also for railway sleepers, harbour pilings and especially fence posts. Deforestation for construction and agriculture greatly reduced the podocarp forests, but the trees are now protected by law; only old fallen logs found outside reserves can be used for timber or carvings. Although slow growing, the tōtara propagates relatively easily and can live to be 1,000 years old. The champion tree, growing in King Country and known as Pouakani, is around 40 metres (130 feet) tall and is reputed to be 1,800 years old.

Today podocarps, along with many species of myrtle, face a new menace from a wind-borne fungus called myrtle rust. It has spread across the Pacific from Brazil, arriving in New Zealand in 2017, and now threatens the survival of many unique native plant species here and in numerous other countries. New Zealand is working in partnership with Kew's Millennium Seed Bank to try to bank as many seeds as possible of this endangered species as an insurance policy for the tree's long-term survival. The remnants of old forests, which once covered huge swathes of New Zealand, are now recovering, and this iconic tree, which has such an ancient lineage and is so closely identified with indigenous culture, may become abundant once again.

FURTHER READING

Ashburner, Kenneth, and Hugh McAllister, *The Genus Betula, A Taxonomic Revision of Birches* (Richmond: Kew Publishing, 2016)

Aughton, Peter, *Endeavour, The Story of Captain Cook's First Great Epic Voyage* (London: Cassell, 2002)

Bail, Murray, *Eucalyptus* (London: Vintage Publishing, 1998)

Bain, Donald, *Explore the Methuselah Grove* (Nova Online, 2001)

Barwick, Margaret, *Tropical and Subtropical Trees. A Worldwide Encyclopaedic Guide* (Portland, OR: Timber Press/London: Thames & Hudson, 2004)

Bean, William Jackson, *Trees and Shrubs Hardy in the British Isles* (London: John Murray, 1950)

Bowett, Adam, *Woods in British Furniture-making 1400–1900. An Illustrated Historical Dictionary* (Wetherby: Oblong Creative Ltd/ RBG Kew, 2012)

Briggs, Gertrude, *A Brief History of Trees* (London: Max Press, 2016)

Brooker, Ian, and David Kleinig, *Eucalyptus: An Illustrated Guide to Identification* (Chatswood, NSW: New Holland Publishers, 2012)

Brooker, Ian, and David Kleinig, *Field Guide to the Eucalypts, Vol.1 South-eastern Australia* (Melbourne and Sydney: Bloomings Books, 1999)

Buchholz, J. T., 'The Distribution, Morphology and Classification of Taiwania (Cupressaceae): An Unpublished Manuscript (1941)', *Taiwania International Journal of Life Sciences*, 58 (2), 2013, 85–103

Bynum, Helen and William, *Remarkable Plants That Shape Our World* (London: Thames & Hudson/Chicago: University of Chicago Press, 2014)

Carey, Frances, *The Tree: Meaning and Myth* (London: British Museum Press, 2012)

Christenhusz, M., M. Fay, and M. Chase, *Plants of the World* (Richmond: Kew Publishing/Chicago: University of Chicago Press, 2017)

Crane, Peter, *Ginkgo* (New Haven and London: Yale University Press, 2013)

Davidson, Alan, *The Oxford Companion to Food* (Oxford: Oxford University Press, 2006)

Desmond, Ray, *The History of the Royal Botanic Gardens, Kew* (Richmond: Kew Publishing, 2007)

Dirr, Michael, *Manual of Woody Landscape Plants* (Champaign, IL: Stipes Publishing Company, 1990)

Dransfield, John, N. W. Uhl, and C. B. Amundson, *Genera Palmarum: The Evolution and Classification of Palms* (Richmond: Kew Publishing, 2nd ed., 2008)

Drori, Jonathan, *Around the World in 80 Trees* (London: Lawrence King Publishing, 2018)

Evarts, John, and Marjorie Popper (eds), *Coast Redwood: A Natural and Cultural History* (Los Olivos, CA: Cachuma Press, 2001, revised 2011)

Farjon, Aljos, *World Checklist and Bibliography of Conifers* (Richmond: Kew Publishing, 2001)

Farjon, Aljos, *A Natural History of Conifers* (Portland, OR: Timber Press, 2008)

Farjon, Aljos, *Ancient Oaks in the English Landscape* (Richmond: Kew Publishing, 2017)

Flanagan, Mark, and Tony Kirkham, *Wilson's China: A Century On* (Richmond: Kew Publishing, 2009)

Flanagan, Mark, and Tony Kirkham, *Plants from the Edge of the World: New Explorations in the Far East* (Portland, OR: Timber Press, 2005)

Fry, Carolyn, *The Plant Hunters: The Adventures of the World's Greatest Botanical Explorers* (London: Andre Deutsch, 2017)

Fry, Janis, *The God Tree* (Milverton: Capall Bann Publishing, 2012)

Gardner, Martin, Paulina Hechenleitner Vega, and Josefina Hepp Castillo, *Plants from the Woods and Forests of Chile* (Edinburgh: Royal Botanic Garden, Edinburgh, 2015)

Gittlen, William, *Discovered Alive: The Story of the Chinese Redwood* (Berkeley, CA: Pierside Publishing, 1999)

Grant, Michael C., 'The Trembling Giant', *Discover Magazine*, October 1993

Grimshaw, John, 'Tree of the Year: *Taiwania cryptomerioides*', International Dendrology Society Yearbook 2010, 24–57

Grimshaw, John, and Ross Bayton, *New Trees: Recent Introductions to Cultivation* (Richmond: Kew Publishing, 2009)

Hageneder, Fred, *Yew: A History* (Stroud: Sutton Publishing, 2007)

Hall, Tony, *The Immortal Yew* (Richmond: Kew Publishing, 2018)

Harkup, Kathryn, *A is for Arsenic: The Poisons of Agatha Christie* (London: Bloomsbury, 2016)

Harmer, Ralph, *Restoration of Neglected Hazel Coppice* (Forest Research Information Note, 2004)

Harrison, Christina, and Lauren Gardiner, *Bizarre Botany* (Richmond: Kew Publishing, 2016)

Harrison, Christina, Martyn Rix, and Masumi Yamanaka, *Treasured Trees* (Richmond: Kew Publishing, 2015)

Hillier, John, *The Hillier Manual of Trees and Shrubs* (Newton Abbott: David & Charles, 1998)

Hogarth, Peter J., *The Biology of Mangroves and Seagrasses* (Oxford: Oxford University Press, 3rd ed., 2015)

Honigsbaum, Mark, *The Fever Trail. In Search of the Cure for Malaria* (London: Macmillan, 2001/New York: Farrar, Straus & Giroux, 2002)

Johnson, Owen, *Tree Register of the British Isles, TROBI* (London: Kew Publishing, 2003)

Johnson, Owen, *Arboretum: A History of the Trees Grown in Britain and Ireland* (Stansted: Whittet Books, 2015)

Lancaster, Roy, *The Hillier Manual of Trees and Shrubs* (London: Royal Horticultural Society, 8th ed., 2014)

Lanner, Ronald M., *Conifers of California* (Los Olivos, CA: Cachuma Press, 2002)

Lewington, Anna, and Edward Parker, *Ancient Trees: Trees that Live for a Thousand Years* (London: Batsford, in association with RBG Kew, 2012)

Lonsdale, David, *Ancient and Other Veteran Trees: Further Guidance on Management* (London: The Tree Council, 2013)

Lyle, Susanna, *Vegetables, Herbs and Spices* (London: Frances Lincoln, 2009)

McNamara, William A., 'Three conifers south of the Yangtze' (Quarryhill Botanical Garden, 2005; http://www.quarryhillbg.org/page16.html. Accessed 16 April 2019)

Magdalena, Carlos, *The Plant Messiah. Adventures in Search of the World's Rarest Species* (London: Viking, 2017)

Manniche, Lise, *An Ancient Egyptian Herbal* (London: British Museum Press, 2006)

Miles, Archie, *The British Oak* (London: Constable, 2013)

Mills, Christopher (ed.), *The Botanical Treasury* (Andre Deutsch in association with RBG Kew, 2016)

Milton, Giles, *Nathaniel's Nutmeg: How One Man's Courage Changed the Course of History* (London: Hodder & Stoughton/ New York: Farrar, Straus & Giroux, 1999)

Mitchell, Alan, *Alan Mitchell's Trees of Britain* (London: Collins, 1996)

Mortimer, J. and B., *Trees and Their Bark* (Hamilton, NZ: Taitua Books, 2004)

Musgrave, Toby, Chris Gardiner, and Will Musgrave, *The Plant Hunters, Two Hundred Years of Adventure and Discovery Around the World* (London: Seven Dials, 2000)

Preston, Richard, *The Wild Trees* (London: Penguin/New York: Random House, 2007)

Rix, Martyn, and Roger Phillips, *The Botanical Garden, Volume 1 Trees and Shrubs* (London: Macmillan, 2002)

Short, Philip, *In Pursuit of Plants: Experiences of Nineteenth & Early Twentieth Century Plant Collectors* (Portland, OR: Timber Press, 2004)

Sibley, David Allen, *The Sibley Guide to Trees* (New York: Knopf Doubleday Publishing Group, 2009)

Smith, Paul (ed.), *The Book of Seeds: A Life-Size Guide to Six Hundred Species from Around the World* (London: Ivy Press, 2018)

Spongberg, Stephen, *A Reunion of Trees: The Discovery of Exotic Plants and Their Introduction into North American and European Landscapes* (Cambridge, MA: Harvard University Press, 1990)

Stafford, Fiona, *The Long, Long Life of Trees* (New Haven and London: Yale University Press, 2017)

Stewart, Amy, *Wicked Plants – the A-Z of plants that kill, main, intoxicate and otherwise offend* (Portland, OR: Timber Press, 2010).

Stokes, Jon, and Donald Rodger, *The Heritage Trees of Britain & Northern Ireland* (London: Constable with The Tree Council, 2004)

Tomlinson, P. B., *The Botany of Mangroves* (Cambridge and New York: Cambridge University Press, 2nd ed., 2016)

Van Pelt, Robert, *Forest Giants of the Pacific Coast* (Seattle: University of Washington Press, 2002)

Vaughn, Bill, *Hawthorn: The Tree That Has Nourished, Healed and Inspired Through the Ages* (New Haven and London: Yale University Press, 2015)

White, Lydia, and Peter Gasson, *Mahogany* (Richmond: Kew Publishing, 2008)

Willis, Kathy, and Carolyn Fry, *Plants: From Roots to Riches* (London: John Murray, 2014)

Woodford, James, *The Wollemi Pine: The Incredible Discovery of a Living Fossil from the Age of the Dinosaurs* (Melbourne: The Text Publishing Company, 2005)

SOURCES OF ILLUSTRATIONS

INDEX